The Pilot's
Radio
Communications
Handbook
3rd Edition

TAB
PRACTICAL
FLYING SERIES

Dedication

THE DATE: August 3, 1981.

THE DECISION: To strike or not to strike.

For many of the 14,000 plus controllers across the county, the decision was anything but easy. Finally, 9,800 or so chose to leave their posts. Three thousand chose to remain.

To those who remained, and to their supervisors, this book is dedicated. They made the hard decision, and, by so doing, ensured the safety and continuity of flight operations throughout the nation.

Without their resolve and sense of responsibility, what could have been chaos became only an inconvenience—and, to most of us, a minor one at that. For their tolerance, their courtesy, and especially the sacrifices they have made, we offer our most sincere thanks. Perhaps by dedicating this effort to these ladies and gentlemen who do so much to ensure our physical well-being, those thanks can become just a little more than mere words.

Other Books in the TAB PRACTICAL FLYING SERIES

ABCs of Safe Flying—2nd Edition *by David Frazier*
Aircraft Systems: Understanding Your Airplane *by David A. Lombardo*
The Art of Instrument Flying *by J.R. Williams*
The Aviator's Guide to Flight Planning *by Donald J. Clausing*
Avoiding Common Pilot Errors—An Air Traffic Controller's View *by John Stewart*
The Pilot's Air Traffic Control Handbook *by Paul E. Illman*

The Pilot's
Radio
Communications
Handbook
3rd Edition

Paul E. Illman and Jay Pouzar

TAB BOOKS Inc.
Blue Ridge Summit, PA

Notice: At press time, a major reorganization of FAR Part 91 was pending final FAA approval. In this edition references to sections of the reorganized Part 91 appear in parentheses. These are tentative section numbers which are subject to change prior to final FAA approval.

Excerpts from aeronautical charts and other publications reproduced herein are for illustration only and are *not for use in navigation*.

THIRD EDITION
SECOND PRINTING

Copyright © 1989 by TAB BOOKS Inc.
Earlier editions copyright © 1984 and 1988 by TAB BOOKS Inc.
Printed in the United States of America

Library of Congress Cataloging in Publication Data

Illman, Paul, E.
     The pilot's radio communications handbook / by Paul E. Illman and
Jay Pouzar.—3rd ed.
          p.   cm.
     Includes index.
     ISBN 0-8306-1445-1 (pbk.)
     1. Radio in aeronautics.   I. Pouzar, Jay F.   II. Title.
TL693.I4   1989   89-4359
629.132'51—dc19   CIP

TAB BOOKS Inc. offers software for sale. For information and catalog, please contact
TAB Software Department, Blue Ridge Summit, PA 17294-0850.

Questions regarding the content of this book
should be addressed to:
     Reader Inquiry Branch
     TAB BOOKS, Inc.
     Blue Ridge Summit, PA 17294-0214

Aviation Editor: Carl H. Silverman
Designer: Jaclyn B. Saunders

Contents

Acknowledgments

We are grateful to a number of individuals who assisted in the preparation of the first, second, and third editions of this book. Without their contributions, we would have found it much more difficult to produce what we hope is a practical and useful discussion of pilot radio communications.

Particular thanks go to Thomas W. Williams for the thoroughness with which he reviewed the original manuscript. As a certificated instrument flight instructor and former air traffic controller, Tom assisted in verifying that the sample radio dialogues and the procedures cited conformed to accepted and standard practices.

We also thank the officials who, when the first edition was published in 1984, were responsible for the operation of the Olathe, Kansas, Air Route Traffic Control Center. Some have since retired or have transferred to other facilities, but they include Rex McQueen, then Air Traffic Manager of the Center, Samuel L. Tyson, the Assistant Traffic Manager, and Robert Kaps, Military Liaison Officer. These gentlemen opened the Olathe Center to us for any information we requested. They, and the controllers who patiently answered our questions, added further authenticity to our efforts.

Equally important was the help provided by the Kansas City Flight Service Station (now located in Columbia, Missouri) and its Chief, Wayne Krueger. The same is true of the Kansas City Downtown Tower and its former Chief, John Munshull. Both organizations and their specialists gave willingly of their time and their knowledge, no matter what our requests.

In addition to the authorities who helped us initially, we must express our appreciation to others who assisted us in the second edition. They include Jerry Schmeltz, Accident Prevention Specialist in the Kansas City Flight Standards District Office, and George Short, the current Air Traffic Manager at the Kansas City Downtown Airport. Jerry, like the others, patiently answered our questions and put us in contact with the appropriate FAA personnel who were specialists in their particular air traffic control fields. George and his Tower specialists brought us up to date on airport traffic control procedures, radar systems, and the new tower itself, which became operational in February 1987. And, we also must thank Doug Perkins, Air Traffic Control Specialist at Wichita's Mid-Continent Airport. As Wichita is a recently-established Airport Radar Service Area (ARSA), Doug's contributions to our understanding of these new air traffic control areas was invaluable.

Of particular assistance was Stewart Morris, area manager at Kansas City Center. Stu shared, most generously, materials, experience, and expertise—all of which added first-hand authenticity to our efforts.

And there were others still: Larry Morin and Mike Weber of the McAlester, Oklahoma, Automated Flight Service Station; Nancy Dobson of the Columbia, Missouri, tower (one of the last bastions of non-radar Approach/Departure Control); Gary Rogers of the aeronautical chart section of the National Ocean Service; and Harry Silverstone, in charge of quality assurance at the Washington, D.C., National Airport tower.

Robert Cunningham, owner of Bob Cunningham Photography, Inc., of Kansas City, Missouri, provided some of the photos. Bob holds commercial, instrument, and multiengine ratings and is the Public Affairs Officer of the Shawnee Mission, Kansas, Civil Air Patrol.

For this third edition we are especially grateful to Eugene G. Devlin, Air Traffic Manager at the McAlester AFSS, for his review of the FSS chapter.

Finally, our appreciation goes to the many controllers and pilots, students and professionals alike, who encouraged us to produce this book. Those close to the subject of radio communications stressed the need for such a book. In so doing, they assured us that it would serve as one more contribution to pilot safety and pilot enjoyment in the air. Such observations only increased our enthusiasm for the project.

We must conclude with a disclaimer of sorts. With only a few exceptions, the aircraft N-numbers in the radio communication examples are those of aircraft once owned by one or the other of the co-authors. Since the initial publication of this book, those aircraft have been sold to other parties—and perhaps re-sold. Consequently, should a subsequent owner find his N-number appearing in a communication example, let that not be a reflection of his radio skills or lack thereof. We chose our own aircraft numbers to avoid any such implication then, now, and in the future.

Introduction

The air is filled today with pilots of all levels of experience, knowledge, and training. There are those who learned to fly at some uncontrolled single-strip airport and are still reluctant to venture too far from that uncomplicated harbor. There are those who mastered the private test at a busy controlled airport. There are the old-timers (World War II vintage) who never lost the flying bug but couldn't afford the luxury until their later years. And there are the pros—the airline captains, the executive pilots, the high-time instructors, the commercial pilots who work charters or are involved in some other money-making enterprises.

The list is hardly complete. Suffice it to say that even with exorbitant fuel and maintenance costs, there are a lot of us in the air—some good, some bad, some in between.

What's the common denominator we all share? Probably the critical one is safety—and all that safety implies. Just below safety may be the freedom that piloting your own (or rented) aircraft brings: the freedom to go places with reasonable economy and speed, and the freedom to get from here to there unencumbered by traffic lights, speed traps, and highway nuts who pass you at 85 on a 65-mph interstate. Only the pilot enjoys the freedoms of space, distance, and speed.

But do we really take advantage of the benefits that the Cessna, Cherokee, or Bonanza offer us? Are we locked into the local traffic pattern or limited to a few short trips to an uncontrolled airport because the big ones scare us? Do we venture into controlled areas with doubts and concerns? How many of us use— or know *how* to use—Approach Control, Center, Flight Service Stations? How many of us have even been taught how these facilities operate and what they can do to make our flights safer and more economical? Said another way, how many of us understand radio procedures and have mastered the techniques of pilot communications?

Considering the emphasis placed on pilot training, medical qualifications, aircraft maintenance, operating rules and regulations, and the like, it would seem that at least a somewhat similar emphasis would be directed to pilot radio communication skills. For whatever reasons, such is not the case. The literature on the subject is too sparse, the examples of radio dialogue too few, and explanations of what to say and how to say it too incomplete.

Theoretically, a book on radio communications shouldn't be necessary; the subject should be part and parcel of every pilot's training. Innumerable discussions with controllers, airline pilots, instructors, and ordinary weekend excursionists, however, indicate just the opposite. According to the pros and amateurs alike, the airwaves suffer from communication misuse, non-use, or overuse.

It's not just hems and haws from fledgling pilots that are disturbing; those we expect and can live with. Instead, it's the garbled, disjointed communications, the incomplete communications, the air-consuming hesitations, the rambling, the lack of knowing what to say and how to say it, and in some cases, the failure to communicate *at all* that arouse concerns. These are the symptoms that a void exists—a void in the literature available, the pilot training process, or both.

In the effort to fill the void, this book has two primary purposes: to contribute to increased safety in flight through timely and correctly worded communications, and to equip the student and licensed pilot with the knowledge of radio communications and the various ground facilities so that his or her flight horizons are expanded beyond the local controlled or uncontrolled airport.

Designed primarily, but not exclusively, for the VFR pilot, the book discusses the whole spectrum of radio facilities and communication responsibilites. First there is MULTICOM, where only aircraft-to-aircraft self-announce messages are exchanged. Then comes a similar treatment of UNICOM, followed by Flight Service Stations, Ground Control, Tower, Approach and Departure Control, and the Air Route Traffic Control Centers.

Accompanying the discussion of the various facilities are explanations of what each does, how to determine the proper frequencies, and most important, examples of what the pilot should say to contact each facility, what he should expect to hear, and how he should respond.

This third edition has been revised to include the latest regulations affecting TCAs and Mode C transponders. In a similar vein, it includes other recent air traffic control and organizational changes, such as the consolidation of Flight Service Stations (FSSs) into regional Automated Flight Service Stations (AFSSs). In some cases, however, reference is made to FSSs, such as Emporia, Kansas, Russell, Kansas, or Goodland, Kansas, that have recently been closed. Those closings, however, do not alter the communication procedures that should be followed when operating in or near any airport that still has an on-site FSS.

1

A Case for
Communication Skills

The weather was clear as we were returning to the Kansas City Downtown Airport from an IFR training flight. After Center had handed us off to Approach and we had established contact, another aircraft made its initial call, also to Approach:

Pilot: Clay County TCA Approach Control, this is Cherokee November Four One Nine Six Six. Over.

Approach: *Cherokee Four One Niner Six Six, Kansas City Approach.*

Pilot: Clay County TCA Approach Control, November Four One Nine Six Six is over the Interstate, and I want to land at the big airport. Over.

Approach: *Cherokee Niner Six Six, squawk zero two five two, ident, and stand by.*

Pilot: Clay County TCA Approach Control, November Four One Nine Six Six squawking zero two five two, identing, and standing by. Over.

At this point the controller directed several other aircraft and lined us up for the instrument approach to Downtown. The controller then returned to N41966.

Approach: *Cherokee Niner Six Six, I missed your ident. Please ident again.*

Pilot: Clay County TCA Approach Control, November Four One Nine Six Six squawking zero two five two and identing. Over.

Approach: [After a pause] *Cherokee Niner Six Six, I'm still not receiving your ident. Remain clear of the TCA, and say your present position and altitude.*

Pilot: Clay County TCA Approach Control, I'm still over the Interstate at three thousand five hundred feet, and I want to land at the big Kansas City airport. Over.

Approach: *Cherokee Niner Six Six, which Interstate are you over? There are several in this area.*

Pilot: Clay County TCA Approach Control, November Four One Nine Six Six. I'm not sure which Interstate, but it's near the city. I still want to land at the big airport. Over.

Approach: *Cherokee Niner Six Six, I have not received your ident. Remain clear of the TCA and stand by.*

Instead of doing what he was told, the pilot of 966 launched into an airwaves-monopolizing discourse along these lines:

Pilot: Clay County TCA Approach Control, this is November Four One Nine Six Six. I don't know why you aren't receiving my indent. I just had it worked on, and the mechanic told me it was fine. I've got to land at the big airport because I told Agnes, my wife, I'd pick her and the kids up when they got in from Chicago. What will they think if I'm not there? Over.

Approach: *Cherokee Niner Six Six, Kansas City International is a Group II TCA, and I can't clear you to land unless your transponder is working. I am not receiving your ident, so remain clear of the TCA, and please stand by.*

Continuing to ignore the explicit instructions, 966 rambled on:

Pilot: Clay County TCA Approach Control, November Four One Nine Six Six. I just had the transponder checked because the last time I was here the controller told me to stand by. I did, and the thing didn't work then. The guy at the radio shop said it worked fine, but I'm still having the same trouble. Can't you get me into the big airport? Over.

Throughout all of this, other aircraft (including ours) were trying to get a word in to report positions, get clearances, and the like. But N41966 continued on and on as though he was the only one in the air.

After the last exchange, the controller saw the light. The pilot of 966, bathed in the glow of ignorance, did what he had been told. He entered "0252" in the transponder, pushed the "Ident" button, and placed the switch in the STANDBY position. Of course he wasn't received!

Once the mystery was solved, the pilot, quite unabashed by his display of incompetence, was cleared into Kansas City International—the "big airport."

This is about as accurate an account of the dialogue as is possible to recreate because, of course, we didn't tape the real thing. It's only one incident, and, while unusual in some respects, it's not very different from what pilots and controllers hear every day. All a pilot has to do is listen with a critical ear. Some of the garbage that filters through speaker or headset from air to ground reflects an appalling lack of knowledge that is both unfunny and potentially hazardous to the ignorant pilot and those occupying the same general airspace.

WHY THE PROBLEM?

Who's to blame for the incompetence? Oh, we could probably point a finger at the instructor who eased over the whole subject of communications, but the main thrust of accusation must be directed at the pilot himself. The pilot of N41966 obviously had little interest in the subject. Otherwise, he would have been sure that he knew what he was doing before venturing into a controlled and congested traffic area. At the same time, we can blame him for a consummate egotism that allowed him to enter such an area with so little knowledge.

Pilots such as our friend in N41966 are dangerous because they don't know what they don't know. They are the airman's example of the Peter Principle. They've risen to their level of incompetence. If the flying ability of the gentleman in the left seat of 966 is comparable to his communicating skills, he'll soon be soaring heavenward on his own wings—with his reliable little Cherokee rusting in some cluttered junkyard.

Yes, we can blame the pilot for incompetence, but others also share in the blame. Let's include the CFIs and CFIIs. And let's include the literature—or lack of it—that discusses the subject of radio communications.

Without exaggeration, it's a subject that probably receives the least attention and explanation of all of a pilot's flight training. Even the material currently in print offers little guidance on what to say and how to say it. The *Airman's Information Manual* takes a stab at a few examples, but the examples are limited; some conclude with the almost extinct "Over."

If flight instructors (certainly not all, but entirely too many) fail to teach more than the absolute rudiments of radio procedures, there are probably three reasons:

The instructor isn't too certain about them himself. This should be an unlikely reason, but one co-author of this book, when getting back into flying after nearly 35 years as a groundling, was told by a young CFI to say "This is *November* 1111 Uniform" and always conclude with "Over." (*November* should really only be used by *controllers* when they don't know the type of aircraft. You should use Cessna, Cherokee, Bonanza, etc., in your call sign. *Over* is rarely used, but nevertheless can be useful to indicate the end of a log transmission. Otherwise, it's not necessary.)

If the instructor has accumulated most of his hours flying out of Cowslip Municipal, he probably isn't very confident of radio techniques. His own insecurity results in a superficial coverage of the subject as he preps his eager students for the FAA check rides.

✈ The instructor is a pro as a radio communicator, but teaching the subject takes time—mostly ground time, which is neither very profitable nor exciting. So the student learns barely enough to get by.

✈ The airport is uncontrolled, with perhaps only UNICOM, and no controlled airport is within reasonable flight range. This is a logical reason for not teaching communications in depth, especially if the student plans only to fly on weekends and demonstrate his skills over Aunt Martha's barnyard. A good instructor emphasizes what is *necessary* to know, not what's *nice* to know.

That same instructor, however, must make it very clear that if the student (now private pilot) ever plans to fly to a controlled airport or through a TCA (Terminal Control Area) or go on a cross-country, he must return for a thorough schooling in radio procedures. A fully equipped aircraft and a private pilot license entitle a pilot to land at any TCA airport. However, the hardware and a piece of paper are hardly adequate to ensure the continued well-being of the pilot or the other airmen in his vicinity. It's dangerous to run out of knowledge—but it can happen easily and quickly to untrained pilot. The results can be devastating.

So okay—you've got a license, and you either own or rent a plane. You're a good pilot, confident of your ability. Now, like many of your counterparts, are you going to spend the rest of your flying days avoiding tower-controlled airports or being fearful of using Center (Air Route Traffic Control Center) on a VFR cross-country? If you've been well-trained in radio procedures, a busy airport or getting advisories from Center is neither a challenge nor a concern. Your radio skill makes flying just that much more fun. But if you're untrained or uncertain, you'll probably steer clear of the controlled areas and not bother Center because you think that's for the IFR pilots and the pros who wheel the wide-bodies.

This, of course, is nonsense. Admittedly, Center may not be able to help you on a busy day if you're VFR. A controller also has the right to refuse to give you routine en route advisories or track you on radar if you come across as hesitant, uncertain, and lacking in knowledge. Center controllers might not do this very often, but requests in VFR conditions have been rejected when the pilot was obviously incompetent in the basics of radio communications. Otherwise, Center exists to serve all pilots—from the greenest student to the 30,000-hour airline captain. Besides, the FAA urges us to use this as well as all other facilities available to us.

Let's be careful not to oversimplify the matter of radio procedures. Mastering them takes time and practice. To underscore that point, the FAA's *Instrument Flying Handbook* makes these comments:

> . . . Many students have no serious difficulty in learning basic aircraft control and radio navigation, but stumble through even the simplest radio communications. During the initial phase of training in Air Traffic Control procedures and radiotelephone techniques, some students experience difficulty . . .
>
> . . . Communication is a two-way effort, and the controller expects you to work toward the same level of competence that he strives to achieve. Tape recordings comparing transmissions by professional pilots and inexperienced or inadequately trained general aviation pilots illustrate the need for effective radiotelephone technique. In a typical instance, an airline pilot reported his position in five seconds whereas a private pilot reporting the same fix took four minutes to transmit essentially the same information . . . The novice forgot to tune his radio properly before transmitting, interrupted other transmissions, repeated unnecessary data, forgot other essential information, requested instructions repeatedly, and created the general impression of cockpit disorganization . . .

PRACTICING FOR COMPETENCE

Mastery of the technique starts with knowing what you want to say, what to listen for, how to respond, and when and how to use the mike that spreads your voice throughout the surrounding skies. As in any other field, the initial ingredient of proficiency is knowledge. The trick is to apply that knowledge in a logical sequence so that you can say what you want to say and get off the air. Once the knowledge is acquired, the next step is practice, followed by more practice, until what you know intellectually becomes an ingrained habit.

If you've ever been asked to make a speech, you know that you didn't just get up and talk. You either wrote the entire speech or outlined it, and then practiced it until you had the subject matter, sequence, body language, and voice inflections down pat. The first time around, you were probably a bit nervous. The second time was a little easier. Eventually, if you spoke or lectured enough, you became an old pro.

It's the same thing talking to the ground from an airplane. Knowledge coupled with practice will calm the nerves and conquer whatever mike fright you might have. No matter how green you are, you'll come across as a professional.

There are a couple of ways you can practice. One is to buy an inexpensive aircraft-band radio that will pick up the various aviation frequencies. Then monitor the transmissions from your home. This will be less effective if you live far from a tower, but you can at least listen to the pilot's side of the communications exchange.

Another practice method is to use a tape recorder and do a little role-playing with yourself. You're the pilot and controller all in one. Make the initial call to Ground Control, and answer yourself as the controller would. Or pretend that you're in flight and want to land at X Airport. Go through the same process. Using your knowledge of radio procedures, act out a series of scenarios on a mythical flight from the first contact with Ground Control until you have "landed," are off the active, and have called Ground Control again for taxi clearance.

Then play the tape back. Be your own worst critic. Be objective about the "dialogue." Ask yourself: "If I were a controller or another pilot listening to me, what would be my impressions of me?" If you're not satisfied, pick up the tape recorder mike, and go at it again.

If you practice this way enough, it won't take long to learn how to get the message across in the fewest possible words and with maximum clarity. Yes, you may feel a little silly sitting there talking to yourself, but that, too, shall pass. Even if it doesn't, it's a small price to pay for greater confidence and increased expertise.

Now you have the words down, but will you remember them, and in the proper sequence, when it comes to the real thing? If in doubt, write out what you want to say when you contact the various services. Put the notes on your knee pad and, if necessary, read from them as you make your calls. After all, we use checklists so that we don't have to rely on a memory that might fail us, so why not adopt the same technique for radio contacts? (However, unlike checklists, experience will make the need for written notes unnecessary.)

A pause to regroup: Are we exaggerating the case and the need for greater communication skills among the pilot population? Obviously, we don't think so. All you have to do is fly a few hours a week and keep an alert ear to what flows through the headset. On any given flight around a busy airport you'll hear everything from a terse "Okay" to a rambling recitation of superfluous trivia to a series of mumbled incoherencies that no human or electronic decoder could decipher. If you question that statement, spend a few minutes with a controller, and listen to what he has to say.

CONTROLLERS ARE HUMAN, TOO

Flight Service Stations, Towers, Approach and Departure Control, and the Air Route Traffic Control Centers are the pilot's valuable but unseen friends. They exist to serve the pilot and to make flying safer for all of us. Their services, however, aren't really free. You've paid for them through your taxes. The services are there to be used or not used, so why not take advantage of your annual donation to Uncle Sam and the airway tax you find tagged onto your fuel bill?

Yes, you've paid for the service, but so has every pilot, so it's not yours and yours alone to use or misuse. Any given service, particularly that offered by Center, can be denied you if you give an impression of incompetence. There are occasions when those on the ground just don't have time to try to make sense out of nonsense and clarity out of obscurity. To do so might put someone else's life in jeopardy.

At times it may seem unlikely, but controllers happen to be human beings too. They have good days; they have bad days. Even on their good days, though, they can quickly turn into vocal ogres when they encounter unmitigated stupidity over the air. On their bad days, they can come across as halo-endowed saints when a knowledgeable pro solicits their assistance or advice.

While we're on the subject, the basic rules of courtesy over the air should *always* prevail. When either the pilot or controller resorts to sarcasm or needless abuse, he is merely reflecting his emotional immaturity—which is hardly a credit to the responsibility he bears. The controllers call it "chipping," a gentle term for "telling the other guy off."

Controllers recognize the humanity of man, and 99 percent never utter a word of recrimination when mistakes are made or ignorance shines brightly. A few don't have that level of patience, of course. They can chip with the best of them, as did one we overheard when he couldn't get a response from a pilot with whom he had just been in contact: "You gonna talk to me, boy? If you are, talk *now*."

To give them their due, controllers have to be models of tolerance and self-control to endure some of the things that go on in and over the air. Yes, some talk too rapidly, and there are those whoruntheirwordstogether so that comprehension is nigh impossible. But the performance of the vast majority, even under pressure, sets a standard of excellence in their profession that pilots should strive to match in *theirs*.

If you have a problem with a controller, don't let anger overrule good judgment. The radio isn't the place for chipping. Wait until you're on the ground. Then call the facility and talk to the supervisor. Explain calmly what happened. Let the supervisor take it from there. Childish spleen-venting is out of place in the adult world—whether airborne or ground-bound.

While the controller is indeed the "controller," that doesn't mean he has to be obeyed at all costs. *You* are still the pilot in command. If the controller tells you to do something that you believe will endanger you, tell him. Don't follow him blindly into the path of possible destruction, but don't keep him in the dark about your concern or your alternate action.

In a very literal sense, a team is at work—you and the person on the ground. He is there to ensure your safety and that of your fellow pilots. He can fulfill his responsibility, however, only if you keep him informed and conduct yourself with the skill expected of a licensed pilot—private or ATP.

By the same token, if you help the controller when he asks you to lengthen your downwind leg, make a tight pattern, land long, speed up, slow down, make a high-speed landing runout, or whatever, you'll be functioning as an effective team member. Remember: the controller can do without you, but you can't do without him. New or experienced, it is entirely to your personal benefit to make it easy for the controller to do his job and thus help you do yours. Achieving that objective is a matter of communications—knowing *what* to say, *how* to say it, *when* to say it, and *why* it should be said. It's knowledge plus skill. The two added together equal professionalism.

All evidence that we have found indicates that a strong case can be made for greater pilot communication skills. The reason behind poor communication, whether it's the absence of literature on the subject or instructor reluctance to emphasize it, is secondary. The result is often a pilot's unnecessary fear of the microphone, which in turn tends to restrict his flying activities and limits the airborne adventures to which his license entitles him. The alternate result is unjustified confidence, as embodied by our friend in Cherokee November 41966.

What follows, from MULTICOM to Center, will hopefully reduce your fear and establish *justified* confidence in your ability to communicate as a professional. Whether flying is your vocation or avocation, that should be your objective.

A FEW WORDS ABOUT PHRASEOLOGY

Because we're about to begin illustrating the various radio calls and contacts, we want to be sure that the accepted phraseology is understood. There's nothing difficult about it, but there is a certain standardization that is both accepted and expected. Reasonable variations are, of course, permissible. The examples that follow in this book, however, generally reflect the approved wording and structure.

As to wording, aircraft N-numbers are stated individually, preceded by the aircraft type. Cherokee 1461 Tango is announced as "Cherokee One Four Six One Tango," not "Cherokee Fourteen Sixty-One Tango." "Land Runway 19" is "Land Runway One Niner," not ". . . Nineteen." "Altimeter 29.65" is "Altimeter two niner six five," not ". . . twenty-nine sixty-five." "Heading 270" is "Heading two seven zero," not ". . . two seventy."

In quoting altitudes, controllers will state them in terms of thousands and hundreds: "maintain three thousand five hundred" or "expect seven thousand five hundred in ten minutes." Pilots can (and do) shorten altitude quotations by saying "level at three point five" or "leaving five point five for three point zero" or "over the field at two point three." This sort of verbal shorthand is acceptable from the pilot, but does *not* conform to FAA standards. Consequently, you won't hear controllers using that phraseology, and in the examples we cite from now on, we will attempt to conform to FAA recommendations.

Accordingly, and to be sure that you understand and employ the correct phraseology, all numbers in the simulated dialogues that follow are spelled out.

"Runway 19" will appear as "Runway One Niner" because that's the way it's pronounced. "Heading 240" is stated as "Heading two four zero" and so on.

Depending on the specific reference, decimal points may or may not be included in the quotation. For instance, when citing altimeter settings, the decimal is omitted. A setting of "30.08" is communicated as "altimeter three zero zero eight." On the other hand, the decimal *is* included in references to radio frequencies. "Contact Ground on 121.9" is stated as "Contact Ground on one two one *point* niner" or "Contact Ground, point niner."

One other explanation is apropos at this point, before we get into examples. You will note that at times we use the aircraft's type and full N-number, such as "Cherokee One Four Six One Tango." On other occasions, it's "Cherokee Six One Tango." Why the difference? When making the *initial* contact with *each* controller (Ground Control, Tower, each Center sector, etc.), the type of aircraft should be identified and its *full* N-number given (just in case the controller is handling *another* "Cherokee Six One Tango," which is a distinct possibility in congested areas). You can shorten the call sign to "Cherokee Six One Tango" *after* the controller does. Once the controller abbreviates your call sign, there's no point in giving the complete identification in subsequent calls to the same controller.

Even at uncontrolled airports, the identification process should be the same. Make the type of aircraft you're flying known to others—and hope they extend the same courtesy to you. There's a big difference between landing behind "Zero Zero Zero Zero Alpha" and "*Learjet* Zero Zero Zero Zero Alpha." It would be nice to know that you're trailing a jet rather than a Piper Cub. The wake turbulence of the former can be a bit more challenging.

Yes, there is a fair amount of verbal shorthand that is acceptable but not necessarily correct. Despite this phraseology latitude exercised by pilots (and perhaps even tolerated by the FAA), we have chosen not to take such liberties in the communication examples. The idea is to present the *correct* wording and phrasing. Accepted but unapproved abbreviations can come later, if you so choose.

2

MULTICOM

A few years ago, one of us was driving with a friend down a well-traveled street in Riyadh, Saudi Arabia. Many side roads intersected this main street, but walls or buildings blocked the driver's view, making it impossible to see any cross traffic that might be approaching. Driving in Saudi Arabia is a thrill in itself, but when there are few stop signs or traffic lights and you can't see cars that might jump out at you from the left or the right, extreme caution is the only alternative to your continued physical well-being.

In this case, our driver friend slowed down at every blind intersection and honked his horn. At night, he still used the horn while blinking his lights. These were the signals to alert others that he was there. They were his nonverbal communications signifying his presence as well as his intentions.

In a more sophisticated sense, MULTICOM is akin to the horn and lights of our Saudi driver. At airports where there is no ground-based traffic control or advisory service, no "red or green lights," no "stop signs," MULTICOM provides the aural communications that reveal your presence and your intentions. Just like the Saudi driver, you are transmitting in the blind to anyone who is tuned to your frequency. You don't know if anyone is really listening or is even in the immediate vicinity. But like the driver, you take that added step—just in case.

In its simplest terms, MULTICOM is nothing more than communications between two aircraft, whether in the traffic pattern or flying at altitude along the same route. While it can be comforting and perhaps important to talk to another pilot during a cross-country to exchange weather information and informal PIREPS, the real value of MULTICOM comes to the fore around uncontrolled, non-UNICOM airports. That's when you need to know *who* is there, *where* he is, and *what* he intends to do—just as he needs to know the same about you. MULTICOM provides the vehicle for that exchange of information over a common frequency.

The key here is a *common* frequency. The FAA has thus established what it calls Common Traffic Advisory Frequencies (CTAFs). The CTAF may be MULTICOM, UNICOM, Flight Service, or Tower frequency (when the tower is not in operation). The only source to determine for certain the CTAF at a given airport is the Airport/Facility Directory, because as the "CTAF" acronym is currently not printed on Sectional Charts. At this writing, the FAA is considering adding CTAFs to the Sectional charts, but until this is done you *must* consult the *A/FD*. Just keep in mind that "Common" doesn't mean one universal frequency for *all* airports, but rather a common frequency that all pilots should use for a given airport, depending on what services, if any, are available. And remember, CTAFs do not exist at airports with full-time control towers.

Let's not be naive, however. No matter how clearly and explicitly you transmit your intentions, not all aircraft have radios. Even if they do, their pilots may not be tuned to the MULTICOM frequency. Indeed, they may not have their radios on at all. ("Why bother with such things at Peapatch Municipal? It'll only drain more life out the tubes of my old Mark 12.")

The alternative is obvious. Flying around an uncontrolled airport demands a swivel neck and sharp eyes. To rely solely on blind transmissions is to invite a few thrills or unexpected encounters of the worst kind. Keep in mind, too, that your transmitter might suddenly go on vacation. You think you're broadcasting to all and sundry, but nothing is passing beyond the mouthpiece of your mike. Such failures have happened—and they could happen to *you* anytime, anywhere. Open eyes and constant head-turning are your best defenses against near hits or close misses.

WHY USE MULTICOM?

Why Use MULTICOM? The answer is already self-evident: safety. If you use it, other aircraft in the area might hear you and use it too. Then everyone will be informed about who is doing what, where, and when.

At the same time, don't be dumb. Be willing to back off when judgment so dictates. You've done a good job of advising others of your actions and intentions, but just as you're about to turn on final, you see some guy coming from nowhere on a long straight-in approach. Decision time—do you *assume* he knows you're there? Do you *assume* he'll give way because you're apparently number one to land? Go ahead and *assume*, but it's a dangerous practice. Break the word into its component parts and you get *ass/u/me*. To assume may make an *ass* out of *u* or *me*.

Discretion says give way. In this case, it's better to be number two than arrive at the same point in a dead heat. Piggybacking may be the way to get a space shuttle to Cape Kennedy, but it's not a very comfortable way for a Cherokee and a Cessna to make a landing. That's a "dead heat" in a very literal sense.

A patently obvious observation? If it is so obvious, why do we still have midair collisions, landing accidents, and an excessive number of close calls? Part of the reason may be complacency ("It can't happen to me.") Part could be ignorance or stupidity—or a combination of the two. In some cases, it's nothing but a flagrant disregard for the rights of others. Whatever the reason, the diligent use of MULTICOM will at least reduce the potential of trouble while helping others who may be as concerned about their well-being as you are about yours.

HOW TO KNOW IF YOU SHOULD USE MULTICOM

How do you know whether you have to rely on MULTICOM at a given airport? Simple. Just check the Sectional. If no ground frequency is listed in the information area for that airport, as in FIG. 2-1, then MULTICOM is the answer. Switch to the CTAF frequency of 122.9, and you're in business. Doubters can check the *A/FD* (FIG. 2-2).

HOW TO USE IT

As with any transmission, follow the Navy's admonition about report writing and correspondence: KISS—Keep It Simple, Stupid. Know what you're going to say, say it in plain English, say it clearly, and keep it short. If you've practiced your radio technique, as suggested in the preceding chapter, you know how you sound. You should have learned to speak distinctly, slowly enough to be understood, and with your message conveyed in an organized sequence. If you

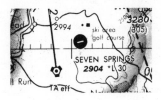

Fig. 2-1. *From the Sectional: A typical MULTICOM airport with no tower, Flight Service Station, or UNICOM.*

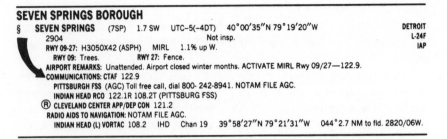

SEVEN SPRINGS BOROUGH
§   **SEVEN SPRINGS**   (7SP)   1.7 SW   UTC–5(–4DT)   40°00′35″N 79°19′20″W   DETROIT
     2904   Not insp.   L-24F
     RWY 09-27: H3050X42 (ASPH)   MIRL   1.1% up W.   IAP
       RWY 09: Trees.   RWY 27: Fence.
     AIRPORT REMARKS: Unattended. Airport closed winter months. ACTIVATE MIRL Rwy 09/27—122.9.
     COMMUNICATIONS: CTAF 122.9
       PITTSBURGH FSS (AGC) Toll free call, dial 800- 242-8941. NOTAM FILE AGC.
       INDIAN HEAD RCO 122.1R 108.2T (PITTSBURG FSS)
     Ⓡ CLEVELAND CENTER APP/DEP CON 121.2
     RADIO AIDS TO NAVIGATION: NOTAM FILE AGC.
       INDIAN HEAD (L) VORTAC 108.2   IHD   Chan 19   39°58′27″N 79°21′31″W   044°2.7 NM to fld. 2820/06W.

Fig. 2-2. *The* Airport/Facility Directory *confirms that the CTAF at Seven Springs is the* MULTICOM *frequency of 122.9.*

speak with the listener in mind, you'll communicate more effectively and with fewer words and less monopolizing of the airwaves.

A SIMULATED LANDING AND DEPARTURE WITH MULTICOM

With the "theory" out of the way, let's talk through the MULTICOM procedures when operating into and out of an uncontrolled airport. Some of the transmissions may seem repetitious, but keep in mind that *safety* is the first concern. A single transmission might be the one that saves the ship.

Approach to the Field

Tune to 122.9 about 10 miles out. If there's any activity, you might get an idea of the traffic volume and pick up the favored runway and wind direction. Assuming this to be the case, start the self-announcing process by identifying your aircraft, position, altitude, and intentions. As there is no control agency of any sort on the field, your call is designed to advise other aircraft of your presence in the area. Consequently, open the transmission with the name of the airport, followed by *"Traffic:"*

> Seven Springs Traffic, Cherokee One Four Six One Tango, ten miles north at five thousand five hundred. Will enter left downwind for Runway Two Seven, full stop Seven Springs.

Note that the name of the airport is repeated at the end of the transmission. This is to make certain that there is no confusion on the part of other aircraft as to the airport to which you are going or at which you are operating. This positive identification is especially important when there are two or three other airports in the general vicinity, each with several aircraft in the air, and all transmitting on 122.9. Identifying the intended airport *twice* will minimize the potential for confusion.

Going back to the arrival example, if you hear nothing after tuning to 122.9, don't be lulled into the belief that the skies are clear. Give yourself every safety edge you can. Announce your intentions to those who might be listening:

> Seven Springs Traffic, Cherokee One Four Six One Tango is ten miles north at four thousand five hundred. Will cross the field at four thousand five hundred for wind tee check, and landing Seven Springs.

Over the Field

Assuming that you've heard no other Seven Springs traffic, you're now over the field and see that the wind tee or sock favors Runway 27. Again, self-announce your intentions to the seen or unseen audience:

> Seven Springs Traffic, Cherokee Six One Tango over the field at four thousand five hundred. Will enter left downwind for Runway Two Seven, full stop, Seven Springs.

Entry to Downwind

You're entering downwind at pattern altitude—approximately 800 feet AGL. Get on the air again:

> Seven Springs Traffic, Cherokee Six One Tango, entering left downwind for Runway Two Seven, full stop, Seven Springs.

Turning Base

Going into the turn from downwind, make your next call:

> Seven Springs Traffic, Cherokee Six One Tango turning left base for Runway Two Seven, full stop, Seven Springs.

At any uncontrolled airport, make this call when *turning* onto the base leg. It's a lot easier for other aircraft to see you when you're in a bank as opposed to straight-and-level flight. Also, a fairly wide pattern with a distinct base leg is preferable to a hotshot U-turn. This gives you time to scan the area for other aircraft on the same leg that haven't announced their presence. It also allows you to check the final approach course for someone who might be making a straight-in approach. This is the altar on which many inflight marriages have been consummated—unwanted but nevertheless eternal marriages.

Turning Final

Once again, announce what you're doing *while you're in the turn* to final:

Seven Springs Traffic, Cherokee Six One Tango turning final, for Runway Two Seven, full stop, Seven Springs.

Clear of the Active

On the ground, get off the runway as quickly but safely as possible. Somebody you never heard of might be on your tail. When clear, don't keep it a secret:

Seven Springs Traffic, Cherokee Six One Tango clear of Runway Two Seven, Seven Springs.

A lot of talk? Yes, but at least you've fulfilled your responsibilities. You've kept others informed, and you've made the air just that much safer. Now if only somebody as considerate is listening…

You've gassed up, coffeed up, and are ready to go again. Once the engine has fired and the radio is on, listen for a moment or two while you're still on the ramp to see if any traffic has developed. As in the pre-landing, your actions can be planned according to what you hear—or don't hear. Regardless, don't assume! Announce your intentions. At most one-strip fields a taxiway is an un-known luxury. Back-taxiing on the active is the only way to get into takeoff position. But let the other guy, if he's out there, know what you're going to do.

Taxi and Back-Taxi

Either while stationary on the ramp or moving slowly toward the runway, make your initial call:

Seven Springs Traffic, Cherokee One Four Six One Tango at south ramp, back-taxiing on Runway Two Seven, Seven Springs.

Now, before venturing onto 27, slow down and scan the approaches for both 9 *and* 27. It's possible that some unheard-from individual is landing downwind or is indeed on the final for 27 but hasn't had the courtesy to inform you.

If the air is clear, start the back-taxi, but don't dawdle. Apply some power and get to the end as rapidly and safely as you can. Otherwise, an approaching aircraft might have to go around. Or worse yet, he might not see you in time to abort his landing—and that could develop into a rather messy situation.

Preflight Check and Runup

Let's assume that there is an area at the end of the runway for the preflight check and engine runup. (If there's not—and you should know this before leaving the ramp or the taxi strip to the runway—complete the check before back-taxiing. Then, when you're at the end, all you have to do is turn around and go.) With an area provided, do a 180 at the end, and park on the *right* side of the departure runway, if you're in the left seat. Then, after the pre-takeoff check has been completed, you can turn toward the runway at a 90-degree angle and have a clear view of any activity on the final approach. Checking for traffic isn't easy if you're in the left seat and are taking the active from the left side of the runway.

This may be a small point, but we've had more than one pilot pull onto the runway when we were on final. He simply hadn't seen us, and a go-around was the only alternative. Was he using MULTICOM? That's a silly question.

Taking the Active

You're ready to go, with the aircraft at an approximately 90-degree angle to the runway. Stop, look upwind and downwind (at this point some pilots swing a complete 360 to observe the entire traffic pattern), and then make your call:

> Seven Springs Traffic, Cherokee One Four Six One Tango departing Runway Two Seven, Seven Springs.

On the matter of someone landing downwind, four conditions could bring about such a landing: (1) very light wind, (2) the pilot's complete disregard of the wind tee or sock, (3) the pilot's failure to monitor or use MULTICOM, or (4) a genuine emergency. All but the fourth are caused by the third. With observant eyes and the radio tuned to 122.9, there is no excuse for aircraft landing in opposite directions at an uncontrolled airport. And yet, because of pilot ignorance, lack of radio equipment, or failure to use the equipment, this sort of thing happens too frequently.

To wit: We recently saw an individual barrel-in downwind, narrowly missing a landing plane that was doing everything right. He discharged his passenger and then roared off from the taxiway-runway intersection, which left him about 2500 feet of a 4000-foot strip. This time, however, he went into the wind. Did this hotshot in his sleek Bonanza use the sophisticated radio equipment that the plethora of antennas implied? Not once. He was apparently above such trivialities.

Departure

Although you've already stated your flight intentions, it doesn't hurt to repeat them once airborne. It's just a courtesy to those still in the pattern:

> Seven Springs Traffic, Cherokee Six One Tango departing Runway Two Seven, departing the pattern to the east, Seven Springs.

Until you're about 10 miles out, stay tuned to 122.9 to pick up any traffic that might be inbound, in your line of flight, or at your altitude. Also, you could help an arriving pilot who has just made his initial call by giving him the wind and runway information. For example:

> Aircraft calling Seven Springs Traffic, Cherokee Six One Tango just departed Seven Springs. Favored runway is Two Seven, winds about two five zero at six.

There's no set pattern for such a call, so help the other pilot in your own words. Use the word "favored," however, when giving runway information. "Active" implies that a particular runway *must* be used—which, at an uncontrolled airport, is not the case.

Touch-and-Gos

Let's suppose that instead of departing the pattern, you want to make a few touch-and-gos. If you're parked at the ramp, the calls are the same up to the time you're ready to take off. Then get on the air:

> Seven Springs Traffic, Cherokee One Four Six One Tango departing [or "taking"] Runway Two Seven, remaining in the pattern [or "closed pattern for touch-and-go"], Seven Springs.

On the downwind, turning base, and turning final, repeat your intentions, just as you did for the full-stop landing:

> Seven Springs Traffic, Cherokee Six One Tango turning downwind for Runway Two Seven, touch-and-go, Seven Springs.

> Seven Springs Traffic, Cherokee Six One Tango turning base for Runway Two Seven, touch-and-go, Seven Springs.

> Seven Springs Traffic, Cherokee Six One Tango turning final for Runway Two Seven, touch-and-go, Seven Springs.

Yes, that's a total of four messages for one touch-and-go, but you may never know who has just entered the pattern. Your last message could be the first he or she has received. You can't be sure—so be safe.

When you've had enough for the day, you're going to land or leave the pattern. In either case, keep the other traffic informed. If it's the final landing, make the downwind, base, and final calls similar to those cited above, substituting "full stop" for "touch-and-go":

> Seven Springs Traffic, Cherokee Six One Tango turning downwind for Runway Two Seven, full stop, Seven Springs.

If you're leaving the pattern, make the call after the last takeoff and when you have the aircraft safely under control:

Seven Springs Traffic, Cherokee Six One Tango departing the pattern to the east, Seven Springs.

CONCLUSION

Inexperienced pilots, pilots who don't know how to maneuver in even moderate traffic, failure to give way to others, lack of knowledge of MULTICOM uses and techniques, failure to turn on the radio, no radio at all—all are reasons for accidents at uncontrolled airports.

In many respects, the controlled airport, even with its five o'clock congestion, creates a greater feeling of security than setting down at a small-town, uncontrolled field. The forced and enforced radio communications make the difference. So at the hundreds of fields like Seven Springs, vigilance coupled with skillful use of MULTICOM will greatly enhance the safety and mental tranquility we all seek in flight.

3

UNICOM

The most modest air-to-ground communication (and also one that can be very helpful) is provided by UNICOM. In a sense, it's a step up from MULTICOM and a step below the tower communications in a controlled airport traffic area.

WHAT IT IS

Simply stated, UNICOM permits radio contact with a ground facility on the airport. At many locations where there is no tower or Flight Service Station, the UNICOM operator fills the void by giving "field advisories" to pilots who call in and request same. The advisory consists of wind direction and velocity, possibly altimeter setting, the favored runway, and any *reported* traffic. Unlike a tower, however, UNICOM is *not* a controlling agency. The operator gives information, and that's all. The rest is up to the pilot.

UNICOM also provides services of a non-flight nature. For example, if you want a fuel truck available for a quick turnaround, the UNICOM operator can make the arrangements. Maybe you need a taxi, or you'd like the operator to call your office or home to advise someone of your arrival time. UNICOM is there to help you.

Of course, if an operating tower or FSS is on the field, UNICOM can't and won't give you runway, winds, or traffic information. That's the responsibility of the official facility. UNICOM will, however, provide the non-flight services mentioned.

WHO OPERATES UNICOM?

At uncontrolled airports (our primary concern at the moment), the fixed-base operator (FBO) usually mans the UNICOM. The radio facility itself can be located anywhere at the airport, but it's generally in the lounge area where a call-in can be handled by the FBO manager or a jack-of-all-trades employee who answers the phone, keeps the books, and sells candy and Sectionals. Typically, a barometer and wind speed/direction indicator are near the transceiver.

Keep in mind that the UNICOM operator is *not* a controller. Indeed, he or she may have only the most meager knowledge of what goes on in the air. He will probably give you the best information he has, but it's unwise to count on 100-percent reliability.

For example, the wind direction and velocity, altimeter setting, and favored runway may be completely accurate. The operator then concludes with "no reported traffic." In reality, a half-dozen planes might be in the pattern, but none has been using the UNICOM frequency for that particular airport. Admittedly, if a half-dozen airplanes are flying around him, it's a bit farfetched to believe that he doesn't know that traffic is in the area; the point, however, is that there has been no radio contact with him or on his frequency. In effect, there's "no reported traffic."

Don't *dis*believe the UNICOM operator, but learn not to depend on his every word. He may simply not be in a position to know everything that's going on outside. Perhaps he's not even a pilot. Or, as in every walk of life, he could be one of the few who just doesn't care. The moral is to use the service, but be vigilant as you enter the airport area. Your own eyes are the best instruments you have to tell you what's happening in the real world.

HOW DO YOU KNOW IF AN AIRPORT HAS UNICOM?

The easiest way is to check the Sectional. Just to be sure there is no confusion about what the Sectional tells in this regard, let's take three examples.

FIGURE 3-1 shows an *uncontrolled* airport (denoted by the magenta symbol and letters) which has only UNICOM. The italicized numbers on the right indicate that 122.8 is the UNICOM frequency, while "L52" reflects the runway length (about 5200 feet) and ±727" the field elevation. "L" is lighted.

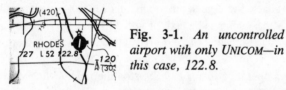

Fig. 3-1. *An uncontrolled airport with only UNICOM—in this case, 122.8.*

FIGURE 3-2 shows an airport with a *Control Zone*. The existence of a Flight Service Station at the airport (as indicated by the heavy-lined box with EMPORIA EMP to the right of the airport symbol) usually creates a Control Zone, as in this case. But there is no tower, so the airport is uncontrolled. The UNICOM frequency, again in slanted numbers, is 122.95. (Don't let some of this throw you. We'll talk more about controlled airports, Airport Traffic Areas, and Control Zones in Chapter 8.)

Fig. 3-2. *An uncontrolled airport within a Control Zone—no tower, but UNICOM and a Flight Service Station are on the field.*

FIGURE 3-3 shows a controlled airport with a tower and a tower frequency of 124.5. How do you know there's a tower? First, because of the letters "CT" (Control Tower) preceding the frequency, and second, the airport symbol is in blue on the Sectional (not distinguishable in this book's black-and-white reproductions). Airports without towers are always depicted in magenta. Those with towers are in blue. UNICOM can be reached on 122.95. You can also find tower, CTAF, and UNICOM information in the *Airport/Facility Directory* (FIG. 3-4).

Keep in mind that at an uncontrolled airport where UNICOM exists, you can get the "full" range of services: wind speed and direction, favored runway, reported traffic, taxicabs, calls to the party of your choice—even the names of good restaurants, if you're curious. At a non-tower airport with a Flight Service Station, the FSS will give you the airport advisory information, but the other amenities come from the UNICOM operator. At a controlled airport with an operating tower, as in FIG. 3-3, the tower provides the aeronautical instructions and information, and you call UNICOM for the ancillary services.

Fig. 3-3. *A controlled airport with a tower and UNICOM but no Flight Service Station on the field.*

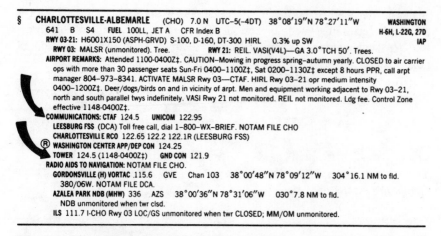

§ **CHARLOTTESVILLE-ALBEMARLE** (CHO) 7.0 N UTC-5(-4DT) 38°08'19"N 78°27'11"W   **WASHINGTON**
   641   B   S4   **FUEL** 100LL, JET A   CFR Index B   **H-6H, L-22G, 27D**
  RWY 03-21: H6001X150 (ASPH-GRVD) S-100, D-160, DT-300 HIRL   0.3% up SW   **IAP**
   RWY 03: MALSR (unmonitored). Tree.   **RWY 21:** REIL. VASI(V4L)—GA 3.0°TCH 50'. Trees.
  AIRPORT REMARKS: Attended 1100-0400Z‡. CAUTION–Mowing in progress spring–autumn yearly. CLOSED to air carrier
   ops with more than 30 passenger seats Sun-Fri 0400–1100Z‡, Sat 0200–1130Z‡ except 8 hours PPR, call arpt
   manager 804–973–8341. ACTIVATE MALSR Rwy 03—CTAF. HIRL Rwy 03–21 opr medium intensity
   0400–1200Z‡. Deer/dogs/birds on and in vicinity of arpt. Men and equipment working adjacent to Rwy 03–21,
   north and south parallel twys indefinitely. VASI Rwy 21 not monitored. REIL not monitored. Ldg fee. Control Zone
   effective 1148-0400Z‡.
  COMMUNICATIONS: CTAF 124.5   **UNICOM** 122.95
   LEESBURG FSS (DCA) Toll free call, dial 1–800–WX–BRIEF. NOTAM FILE CHO
   CHARLOTTESVILLE RCO 122.65 122.2 122.1R (LEESBURG FSS)
  ® **WASHINGTON CENTER APP/DEP CON** 124.25
   TOWER 124.5 (1148-0400Z‡)   **GND CON** 121.9
  RADIO AIDS TO NAVIGATION: NOTAM FILE CHO.
   GORDONSVILLE (H) VORTAC .115.6   GVE   Chan 103   38°00'48"N 78°09'12"W   304°16.1 NM to fld.
   380/06W. NOTAM FILE DCA.
   AZALEA PARK NDB (MHW) 336   AZS   38°00'36"N 78°31'06"W   030°7.8 NM to fld.
   NDB unmonitored when twr clsd.
   ILS 111.7 I-CHO Rwy 03 LOC/GS unmonitored when twr CLOSED; MM/OM unmonitored.

Fig. 3-4. *The A/FD confirms the UNICOM frequency as 122.95, but note that the CTAF—to be used for self-announced traffic advisories when the tower is closed—is the tower frequency of 124.5, not the UNICOM frequency.*

UNICOM's major drawback is that it may not be in operation when you need it. Perhaps the person in charge has gone to lunch. Maybe he or she is gassing an airplane, is on the phone, or just doesn't want to be bothered. The vast majority of operators are conscientious business people who want to render a service; a few, however, simply consider UNICOM an interruption of more profitable pursuits.

The service can thus be limited. Otherwise, it provides another safety dimension that every pilot can and should use.

CONTACTING UNICOM

Recall what we said about CTAFs in the previous chapter. At airports where there is UNICOM *but no tower or FSS*, the UNICOM frequency (printed on the Sectional and in the *A/FD*) is the CTAF and is most often 122.7, 122.8, or 123.0. Let's say that you're going into Rhodes Airport, the example in FIG. 3-1. Accordingly, you tune in 122.8 about 10 miles out. Just as you did with MULTICOM, monitor the frequency to see what you can learn from other aircraft that might be in the pattern or what information UNICOM may be relaying to other aircraft. If you can pick up the winds, favored runway, and so on just by eavesdropping, so much the better. You can spare the air that one transmission.

For now, let's assume that all is silent as you approach Rhodes Airport. About 10 miles out, the initial call to obtain the field advisory should go like this:

You: Rhodes UNICOM, Cherokee One Four Six One Tango.

UNICOM: Cherokee One Four Six One Tango, Rhodes UNICOM.

You: Rhodes, Cherokee Six One Tango is ten miles south at four thousand, landing Rhodes. Request field advisory.

UNICOM: Cherokee Six One Tango, wind two three zero at one zero, variable. Favored runway is One Niner. Altimeter two niner six five. Reported traffic two Cessnas in the pattern.

A word of caution, however: Be sure you're getting the right information from the right field. As just one example, in Kansas, there are three airports—Ottawa, Paola, and Garnett—within 25 miles of each other. All three UNICOMs transmit and receive on 122.8. You don't have to have much altitude to pick up all three, so you could be receiving Paola winds and traffic when you wanted to land at Ottawa, or vice versa. This is why it's desirable to begin and end your traffic advisory messages with the name of the airport, just as you do with MULTICOM.

At this point, depending on your requirements, several options are open to you. All you want to do is land and tie down: "Roger, Rhodes. Cherokee Six One Tango."

You'd like a taxicab: "Roger, Rhodes. Would you call a taxi to take us to the Zandu Products office?"

You need a mechanic: "Roger, Rhodes. Would you have a mechanic available? Our oil temperature is running high."

You'd like to have UNICOM make a phone call: "Roger, Rhodes. Would you call 555-5678 and advise Mr. Schwartz that his party from Metropolis will be about 15 minutes late?"

But what if UNICOM doesn't respond? Try to rouse the operator a couple of more times with "Rhodes UNICOM, Cherokee One Four Six One Tango."

If you still get no response, make a blind call on the CTAF of 122.8 (UNICOM frequency) to any aircraft that might be in the pattern: "Any aircraft at Rhodes Airport, this is Cherokee One Four Six One Tango. Can you give me a field advisory?

If someone answers you, all to the good. You can proceed to enter the pattern. Otherwise, it would be smart to fly over the field for a wind tee or sock check. In this instance, the call is the same as with MULTICOM:

Rhodes *Traffic*, Cherokee One Four Six One Tango is ten miles north at four thousand. Will cross the field at three thousand for wind check, Rhodes.

Note that from this point on, whether you made contact with UNICOM or not, all flight-related calls are made to "Rhodes *Traffic*," not UNICOM. UNICOM has absolutely no control over what takes place out there, so it's incumbent on each pilot to communicate his or her intentions and actions to all other pilots in Hence, *Traffic* is now substituted for UNICOM.

A SIMULATED LANDING
AND DEPARTURE WITH UNICOM

With or without an initial contact with UNICOM, all calls now follow the models we discussed in the MULTICOM chapter. However, to be sure that the routine transmissions are illustrated, let's go through them, but without repeating the various conditions and observations.

Before Entering the Pattern

Case One: You received no field advisory from any source, so you fly over the field to determine the wind direction:

Rhodes Traffic, Cherokee One Four Six One Tango over the field at two thousand five hundred. Will enter left downwind for Runway One Niner, full stop, Rhodes. (2500 feet MSL puts you 1800 feet above the 727-feet MSL field elevation.)

Case Two: You received a field advisory from UNICOM or another aircraft:

Rhodes Traffic, Cherokee One Four Six One Tango entering left downwind for Runway One Niner, full stop, Rhodes.

Despite what we said above about no admonitions, we have to issue this one again: *Keep your eyes open*. Yes, UNICOM said two Cessnas were reported in the pattern, but do you *know* that they constitute the only traffic? Has someone else shown up with no radio, a radio that hasn't been turned on or tuned in, or who hasn't bothered to report his presence? This is an *uncontrolled* airport, so your most effective life preserver may be a healthy skepticism, liberally sprinkled with vigilance.

Turning Base and Final

Rhodes Traffic, Cherokee Six One Tango turning left base for Runway One Niner, full stop, Rhodes.

Rhodes Traffic, Cherokee Six One Tango turning final for Runway One Niner, full stop, Rhodes.

Down and Clear of the Runway

Rhodes Traffic, Cherokee Six One Tango clear of Runway One Niner, Rhodes.

Departure from a UNICOM airport is essentially the same as with MULTICOM, except that a call to UNICOM can give you the winds, favored runway, and traffic information—if you haven't already obtained them from a personal visit with the operator. Assuming that you haven't, the initial contact would go like this:

Rhodes UNICOM, Cherokee One Four Six One Tango on the ramp. Request airport advisory.

Once you have the advisory, and while still stationary or taxiing slowly, get on the air again and tell the local traffic what you're doing (or going to do):

Rhodes Traffic, Cherokee One Four Six One Tango taxiing to [or back-taxiing on] Runway One Niner, Rhodes.

Departure

Rhodes Traffic, Cherokee One Four Six One Tango departing Runway One Niner, departing the pattern to the east, Rhodes.

After Takeoff

Rhodes Traffic, Cherokee Six One Tango departing the pattern to the east, Rhodes.

Touch-and-Gos

As with MULTICOM, communicate your intentions before taking off:

Rhodes Traffic, Cherokee One Four Six One Tango departing Runway One Niner, remaining in the pattern for touch-and-go, Rhodes.

On downwind:

Rhodes Traffic, Cherokee Six One Tango turning downwind for Runway One Niner, touch-and-go, Rhodes.

Turning base:

Rhodes Traffic, Cherokee Six One Tango turning base for Runway One Niner, touch-and-go, Rhodes.

Turning final:

Rhodes Traffic, Cherokee Six One Tango turning final for Runway One Niner, touch-and-go, Rhodes.

Landing or Departing the Pattern After Touch-and-Gos

When you're through practicing and want to land, advise the traffic accordingly on downwind, base, and final:

Rhodes Traffic, Cherokee Six One Tango downwind for Runway One Niner, full stop, Rhodes.

And so on.

If departing the pattern, this message will keep the traffic informed:

Rhodes Traffic, Cherokee Six One Tango departing the pattern to the east, Rhodes.

Now stay tuned to the Rhodes CTAF (UNICOM) until you're 10 miles or so from the airport. If you happen to hear an inbound pilot calling Rhodes UNICOM for a field advisory, *don't* jump in and volunteer the information, as we suggested in the MULTICOM chapter. Let UNICOM provide the information. If, after a number of attempts to raise the UNICOM operator, there is no response, it's then proper to help the other person out, based on what you knew a few minutes ago:

Aircraft calling Rhodes UNICOM, Cherokee Six One Tango just departed Rhodes. Favored runway is One Niner, winds about one niner zero at two zero.

CONCLUSION

Flying around an uncontrolled airport presents many opportunities for unwanted confrontations. Whether MULTICOM or UNICOM, the potential is the same. All we can do is urge caution and compliance with the standard radio procedures.

To quote the FAA:

. . . increased traffic at many uncontrolled airports require[s] the highest degree of vigilance on the part of pilots to see and avoid aircraft while operating to or from such airports. Pilots should stay alert at all times, anticipate the unexpected, use the published CTAF frequency, and follow recommended airport advisory practices.

Whether all by yourself in the pattern or one of several, follow the radio procedures we've outlined. Coupled with a swivel neck and alert eyes, sharp radio technique will make the air a lot safer for everyone—especially you.

4

Flight Service Stations

As close as the phone—or maybe a short walk—is the pilot's supermarket of information and assistance. No, the Flight Service Station can't do everything, but when it comes to flight planning, weather, airport advisories, or *almost* anything, the FSS is a storehouse of aids that pilots need, use, or should use. If you're merely going to shoot a few touch-and-gos at the local aerodrome or wander around within a 25-mile radius of home base, the need for FSS services may be limited. But a venture further from home demands at least a brief phone call.

For example, you're going to a field 30 miles or so away. A gusty crosswind blows you off the runway on landing and you wipe out a gear or catch a wingtip. Did you check the weather before taking off, and was your N-number recorded? If not, you might have to foot the repair bill out of your own pocket. Some aircraft insurance policies are invalid unless you can substantiate that you had taken that simple preflight precaution. The FSS was there, but you didn't use it.

The Services Offered: In Summary

An FSS can offer you:

- Weather information—local, en route, and terminal, including sky conditions, winds, temperatures, dewpoints, icing, frontal activity, trends—you name it, the FSS has it.

☒ PIREPs (Pilot Reports) of conditions not easily determined from available charts or data.

☒ Flight plans—filing, changing, extending, and closing.

☒ Airport advisories for the airport where the FSS is located (when there is no tower on the airport or the tower is closed).

☒ Airport advisories (excluding traffic information) for airports with no FSS, but having a Remote Communications Outlet (RCO) and a weather observer.

☒ Status of Restricted Areas and Military Operations Areas (MOAs).

☒ NOTAMs—Notices To Airmen regarding airport conditions, hazards, etc.

☒ Emergency assistance services—direction finding (DF) fixes, steers, and approaches; VOR and ADF orientations; etc.

Add one more. a sense of security, if you want to call it that. Thirty minutes after your estimated time of arrival, if you haven't extended or closed your flight plan, the FSS is on the phone checking your whereabouts. It's comforting to know that somebody down there is watching you.

With the library of help and information the FSS can offer, let's start at the beginning and follow a rough sequence of how you might make use of the various services.

PREFLIGHT PLANNING INFORMATION

Except when contacting the AFSS (Automated Flight Service Station), the considerate pilot makes *two* preflight calls or visits to the FSS: the first to get the necessary information so the flight can be planned accordingly, and the second to file the plan tersely and in the proper sequence. Don't take up the time of FSS personnel, whether by phone or in person, accumulating the data, computing your heading, determining ground speed, elapsed time, and the rest, and then filing the plan. Get the information first, then call back when you know what you're talking about. This is a mutually time-conserving courtesy, but not all pilots follow it.

An in-person visit to the FSS, when possible, almost always produces a helpful dialogue between pilot and specialist. Such visits have become less convenient, however, because the FAA is consolidating the older local FSSs into 61 regional Automated Flight Service Stations. All AFSSs will be in operation by 1990, and the remaining older stations will close by 1994. And even if you have an FSS at your airport, it might not be easily accessible from your location on the field. When possible, though, try to get a briefing in person. It's much more meaningful to read and study the Area Forecasts, Surface Analysis Charts, Weather Depiction Charts, PIREPs, NOTAMs, and the rest than to try to glean needed information over the telephone. A picture is worth . . . well . . .

Conversely, the telephone is a somewhat impersonal communication medium which can result in misunderstanding, confusion, or insufficient information. The vast majority of Flight Service specialists give thorough weather briefings, but some only answer the questions asked. And in busy periods when Instrument Meteorological Conditions exist over the FSS's service area, the specialist may be too busy to provide an in-depth briefing for the VFR customer. Particularly over the phone, prepare in advance all of the questions you want answered, give the specialist the information he needs at the start, and be ready to copy. As in the air, don't ramble on and consume the specialist's valuable time—as this character did:

FSS:	*Good morning. Lake City Flight Service.*
Pilot:	Hi. I'm trying to get to Mountain Town. How are things between here and there? [Let's say Mountain Town is 300 NM from Lake City.]
FSS:	*When do you plan to leave, sir?*
Pilot:	Oh, in about an hour—say ten o'clock.
FSS:	*Will you be VFR?*
Pilot:	That is roger. [That shows he's got the lingo down.]
FSS:	*And at what altitude do you intend to fly?*
Pilot:	I guess that depends on what you tell me about the winds. How're they running?
FSS:	*Well, let's see. At 3000 they're at 290 at 20; 6000, 240 at 30; 9000, 220 at 35—generally southwest to west.*
Pilot:	And what's the ceiling? CAVU all the way? [That's an I-know-what-I'm-talking-about acronym for "Ceiling and Visibility Unlimited."]
FSS:	*No. Lake City is reporting 3000 scattered, 8000 broken, Midpoint Junction is 2500 broken, and Mountain Town is 4000 overcast.*
Pilot:	Kinda marginal VFR in spots, isn't it?
FSS:	*Yes, sir.*
Pilot:	What's the forecast for later today?
FSS:	*About the same conditions until 2100 Zulu.*
Pilot:	2100 Zulu—that's three o'clock tomorrow . . .
FSS:	*No, sir, 2100 is 1500 local; and 1500 local is 3:00 P.M. today.*
Pilot:	Oh, yeah. I always get confused about whether to add or subtract. Well, what else can you tell me?
FSS:	*. . . Excuse me, sir. I have another call. Stand by.*

If the specialist is lucky, a telephone failure will now ensue.

If you call an FSS for a briefing, be informative and to the point. Speak slowly and clearly to allow the specialist to absorb and digest the information as he types it into his terminal. This way he can tell you what you want to know without a lot of needless questions. Plan your call, and be sure to give the following information:

✈ VFR or IFR

✈ Your N number or your name (if you don't know which plane you will be flying)

✈ Aircraft type

✈ Point of departure

✈ Route of flight

✈ Destination

✈ Desired altitude

✈ Estimated departure time and estimated time en route

✈ Type of briefing desired: Standard, Abbreviated, or Outlook (see the *AIM* for details)

And then be prepared to copy. Save your questions until he's done.

Now, hang up, get your Avstar or E6B, and plan the flight. Once the paperwork is done, record it on an FAA Flight Plan form, call the FSS back, and read off the information. Then be sure to get the proper frequency so you can contact the Station when you're ready to open the flight plan. If, for any reason, you fail to get the proper frequency (or forget it), don't hesitate to contact the FSS over any VOR voice facility, RCO, or other published non-emergency frequency.

Many FSSs now offer Fast File, a service by which you can record your flight plan by phone without speaking to a specialist. This service is handy on busy days, but is not the preferred method of filing, because the FSS may be unable to reach you if they need a clairification of your routing, etc.

The typical Flight Service Station is a fairly busy place. The specialists have no time to participate in idle chatter or drag information from an unprepared airman. Know what you want, ask for it, and get off the phone. You're dealing with professionals, so be professional yourself.

OPENING THE FLIGHT PLAN

If the Flight Service Station is on the field of your departure, contact it on the frequency given when you filed the flight plan. To simplify matters, make the contact after the pre-takeoff runup rather than after takeoff when you may

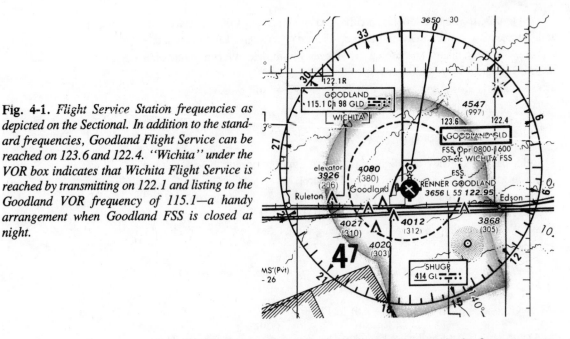

Fig. 4-1. *Flight Service Station frequencies as depicted on the Sectional. In addition to the standard frequencies, Goodland Flight Service can be reached on 123.6 and 122.4. "Wichita" under the VOR box indicates that Wichita Flight Service is reached by transmitting on 122.1 and listing to the Goodland VOR frequency of 115.1—a handy arrangement when Goodland FSS is closed at night.*

be a bit busy. In this and all other calls to an FSS, identify the frequency you are monitoring. Doing so helps the specialist key to the right frequency and respond more promptly.

How do you know what frequency to use if you aren't told in advance? Take Goodland FSS in FIG. 4-1 for example. Above the bold FSS box, you see "123.6" and "122.4". Remember, though, that 123.6 is reserved for airport advisories at those fields where there is an FSS but no tower in operation.

Note that 121.5 (which is for emergency use only), 122.2, 243.0, and 255.4 are never indicated above any FSS box. This is because they are universal frequencies—available at every Flight Service Station. So if you have a standard VHF radio, you could contact Goodland on any of the following: 122.2, 122.4, 123.6, and, in an emergency situation, 121.5.

Also note that Goodland FSS, like most of the remaining non-automated FSSs, only operates part-time (in this case, from 0800-1600 hours local time). At other times you would contact Wichita FSS, via the Goodland VOR. To do this, transmit on 122.1 (the "R" next to the frequency), and listen on the VOR frequency of 115.1.

When opening the flight plan, remember the frequency the FSS told you to use and identify it in the initial call:

Goodland Radio, Cherokee One Four Six One Tango on one two two point two.

And, as we said above, this principle applies to any call to Flight Service, regardless of purpose.

Note this, too, in the Goodland call: You're contacting Flight Service, but you address the call to "Goodland *Radio*"—not "Goodland Flight Service." "Radio" is the standard terminology whenever you initiate any communication with a Flight Service Station.

If Goodland is closed you would call Wichita FSS as follows:

Wichita Radio, Cherokee One Four Six One Tango, listening Goodland VOR. [In this case, they'll know you're on 122.1 because that's the only frequency you can use to transmit over the VOR].

But back to opening the flight plan and an important point: If you are at a *controlled* airport with a tower and Ground Control, always advise Ground Control before you leave Ground's frequency for any frequency other than Tower. Until you're ready to depart and require clearance from Tower, you're under Ground Control's jurisdiction. So, before contacting Flight Service, ask Ground for permission:

Lake City Ground, Cherokee Six One Tango, request frequency change to Flight Service.

After Ground approves, make the call to Flight Service:

You:	Lake City Radio, Cherokee One Four Six One Tango on one two two point six.
FSS:	*Cherokee One Four Six One Tango, Lake City Radio.*
You:	Lake City, Cherokee One Four Six One Tango. Please open my flight plan to Mountain Town at this time.
FSS:	*Cherokee Six One Tango, roger. We will open your flight plan at one zero.* [Meaning "ten minutes past the hour."]
You:	Roger. Thank you. Cherokee Six One Tango.

Now go back to Ground and reestablish communications:

Lake City Ground, Cherokee One Four Six One Tango is back with you.

We won't go through the succeeding contacts with the tower, as those are covered in a later chapter. Let's just say that if you're off on time, there's nothing more to do at this point as far as Flight Service is concerned. On the other hand, should an unexpected delay occur, keep the FSS advised and revise your departure time accordingly.

If the Flight Service Station is *not* physically located on the field of your departure, what do you do then? For the sake of illustration, let's take Lucas, Kansas. Lucas is served by the Russell FSS, approximately 20 miles to the southwest, and is identified on the Sectional in FIG. 4-2.

Assuming that Russell told you to call them on 122.6 when you filed the flight plan, make the initial contact when you're airborne and have sufficient

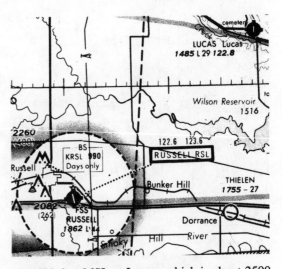

Fig. 4-2. *You'll need to climb a bit to reach the Russell FSS by radio from over Lucas.*

altitude for radio reception—say 4000 feet MSL at Lucas, which is about 2500 feet AGL. As on the ground, the call is much the same:

You:	Russell Radio, Cherokee One Four Six One Tango, on one two two point six.
FSS:	*Cherokee One Four Six One Tango, Russell Radio.*
You:	Russell, Cherokee One Four Six One Tango is off Lucas at four seven. Would you open my flight plan to Wichita at this time?
FSS:	*Cherokee Six One Tango, roger. We have you off Lucas at four seven and will open your flight plan at this time.*
You:	Roger. Thank you. Cherokee Six One Tango.

One other situation: you're at a field that has a Remote Communications Outlet (RCO), such as Hays, Kansas, 25 NM west of Russell (FIG. 4-3). The FSS is in Russell, but the remoted outlet at Hays extends its service range. From a radio contact point of view, it's as though the FSS were located at Hays.

In cases such as this, calls to open or close a flight plan can be made on the ground—no need to get altitude beforehand. Just call:

Russell Radio, Cherokee One Four Six One Tango, on Hays, one two two point three.

Or, although not the "recommended" phraseology, you can call:

Hays Radio, Cherokee One Four Six One Tango on one two two point three.

Although there is no such thing as Hays FSS, and therefore no "Hays Radio," the specialists at Russell FSS will respond nonetheless. As more FSSs are

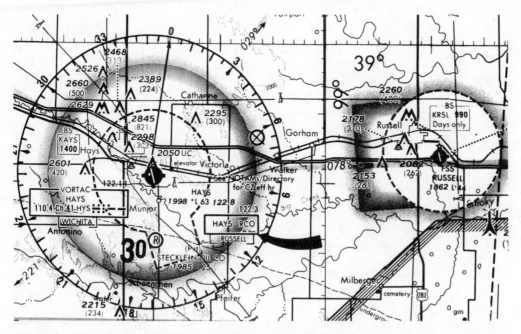

Fig. 4-3. *Although the Flight Service Station is in Russell, Hays has a Remote Communications Outlet (RCO) on 122.3, which permits radio contacts with the FSS as though it were on the airport.*

consolidated the greater the importance of announcing both the facility name and the frequency will be. A specialist nowadays might be handling dozens of frequencies and RCOs at once. And more than one RCO might use the same frequency. For example, the McAlester (Oklahoma) Automated FSS handles RCOs in Stillwater and Oklahoma City, both of which operate on 122.4. The more specific you are in your callup, the faster the response will be—especially when the specialist's attention is diverted away from his console. RCO coverage generally extends to 4000 feet AGL at 50 NM.

FILING A FLIGHT PLAN IN THE AIR

Whether VFR or IFR, filing a flight plan in the air is not the thing to do—unless there is no other alternative. Let's assume, however, that once you get airborne on your cross-country, better judgment prevails and you decide that it would be nice if some people down there knew who you were and where you were going.

The flight is nonstop, Dalhart, Texas, to Wichita, Kansas. Approaching the Liberal VOR, you work out the required details, write them down on the FAA Flight Plan form (FIG. 4-4), and then contact the Wichita Flight Service by transmitting on 121.1 while listening on the 112.3 VOR frequency.

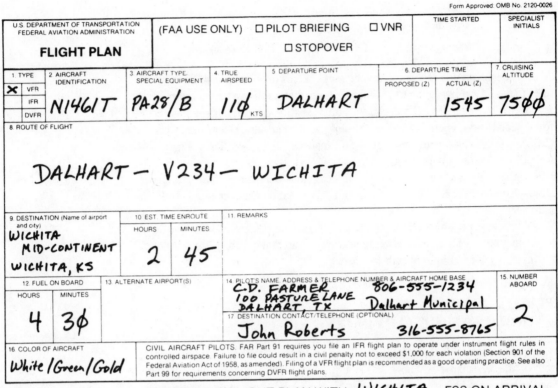

Fig. 4-4. *An example of a properly completed VFR flight plan. If filed on the ground, proposed departure time would be included.*

You: Wichita Radio, Cherokee One Four Six One Tango, listening Liberal VOR.

FSS: *Cherokee One Four Six One Tango, Wichita Radio.*

You: Cherokee Six One Tango requests air filing of a VFR flight plan. Advise when ready to copy.

FSS: *Cherokee Six One Tango, go ahead.*

You: Roger. VFR—November One Four Six One Tango—PA Two Eight slash B—one one zero—off Dalhart at four five—seven thousand five hundred—direct via Victor Two Three Four—Wichita Mid-Continent—two plus four five—four plus three zero—C.D. Farmer, One Zero Zero Pasture Lane, Dalhart, Texas, eight zero six dash five five five dash one two three four—Dalhart Municipal—two—white, green, gold—John Roberts, three one six dash five five five dash eight seven six five.

> FSS: *Cherokee Six One Tango, Roger. We show you off at four five. Liberal altimeter three zero zero six.*
>
> You: Three zero zero six. Cherokee Six One Tango. Thank you.

If the information relayed in this message seems disjointed or to be in code, refer to the FAA Flight Plan form and you'll see that it follows the recommended sequence: Type of flight (VFR)—Aircraft Identification—Aircraft Type slash code for Special Equipment—Airspeed—Point of Departure—Departure Time—Cruising Altitude—Route of Flight—Destination—Estimated Time Enroute—Remarks (none in this illustration)—Fuel on Board—Alternate Airport(s) (For IFR flights)—Pilot's Name, Address, and Telephone Number and Aircraft Home Base—Number Aboard—Color of Aircraft—Destination Contact and Telephone Number.

While we have illustrated filing via a VOR, obviously, the same can be done by contacting an FSS direct, assuming you are within range. Filing is never done, however, over the Flight Watch frequency.

FLIGHT WATCH/
EN ROUTE FLIGHT ADVISORY SERVICE

You're on a VFR flight from Goodland, Kansas (near the Colorado border) to Wichita, a distance of about 250 statute miles. It's springtime in the Plains states, and around Hays you notice a fair-sized buildup of cumulus activity ahead of you and to the southeast. Alert to the potential ferocity of thunderheads in this part of the country, you think a little information about what is happening out yonder would be helpful.

You could call the Wichita FSS over the Hays VOR or over the Hays RCO. But if you're seeking *only* weather information for your route, it's best to call Flight Watch instead.

Flight Watch is an en route flight advisory service *only*. You cannot use it for preflight briefings, or opening or closing a flight plan. Simply and solely, it is a weather information position in the FSS designed to provide the pilot with specific information about his specific route of flight.

You can reach Flight Watch from anywhere in the U.S. if you are at least 5000 feet AGL. And the frequency is *always* 122.0 unless you're flying above 18,000 feet MSL. (Above that altitude, each Air Route Traffic Control Center area will have its own discrete Flight Watch frequency.) Flight Watch operates from 6 A.M. to 10 P.M. local time, daily.

The existence of Flight Watch is not indicated on the Sectionals any longer, because the service is now available nationwide. So how do you know which Flight Watch station to call? There's a chart on the inside back cover of the *A/FD.*

You:	Wichita Flight Watch, Cherokee One Four Six One Tango, Hays VOR.
FW:	*Cherokee One Four Six One Tango, Wichita Flight Watch.*
You:	Flight Watch, Cherokee One Four Six One Tango is VFR on Victor One Three Two, 20 miles south of Hays en route Wichita at seven thousand five hundred. Request en route weather advisories.
FW:	*Cherokee Six One Tango, cumulus building east of your route, with potential thunderstorms northeast of Gage. Wichita is clear at this time, but southwesterly winds expected to increase to two zero to three zero knots by one six three zero local. No PIREPs have been received and no SIGMETs issued up to now over your route.*
You:	Roger, Flight Watch. Thank you. Cherokee Six One Tango.

Just dial in 122.0, call ''Flight Watch'' and indicate the nearest VOR or your position. The appropriate Flight Watch (FSS) will answer.

There's a good chance that the Flight Watch Specialist will solicit a PIREP from you. Report the favorable conditions as well as the bad and advise him of your aircraft type.

Another alternative for en route weather briefings, of course, is to tune to a VOR or NDB that provides Transcribed Weather Broadcasts (TWEB). These are continuous transmissions covering the weather within a 400-mile radius from the station. This is not, however, a service offered by every VOR or NDB. The stations that broadcast TWEBs are identified by a small square in the lower right corner of the identifier box (FIG. 4.5).

Fig. 4-5. *Transcribed Weather Broadcast (TWEB) service over a navaid is indicated by the solid square in the lower right corner of the identifier box.*

IN-FLIGHT WEATHER ADVISORIES

The purpose of this book is *not* to indulge in a meteorological discussion. The dangers of icing, fog, thunderstorms, turbulence, low ceilings, wind shears, and other weather-related phenomena are—or should be—well-known. What is not as thoroughly understood by the weekend or lightplane VFR pilot is the in-flight weather information provided by Flight Service that updates en route conditions and alerts pilots to conditions that might be encountered. Above and beyond Flight Watch, TWEBs, and so on, there are Severe Weather Forecast Alerts, AIRMETs, SIGMETs, Convective SIGMETs, and PIREPs, each of which offers information that may be vital to the pilot.

Severe Weather Forecast Alerts are issued to advise of a a pending Severe

Weather Bulletin regarding severe thunderstorms and tornado activity.

AIRMETs reflect conditions that present a potential hazard to lightplanes and to those lacking instrumentation or equipment (as deicers or anti-icers). In effect, an AIRMET is an amendment to the area forecast.

SIGMETs report significant meteorological developments that could be particularly hazardous to light aircraft and potentially hazardous to all aircraft. They will also be included in the area forecast.

All Flight Service Stations within 150 miles of the weather area broadcast SIGMETs upon issuance and at 15 and 45 minutes past the hour during the first hour. Thereafter, an alert notice is broadcast at 15 and 45 minutes past the hour (for the duration of the advisory), stating that "SIGMET (name and number) is current."

If you're monitoring an FSS or a VOR and pick up only the alert notice (not having heard the initial and complete SIGMET or AIRMET), you would be wise to call either the FSS or Flight Watch for further information:

You:	Shreveport Radio, Cherokee One Four Six One Tango on one two two point two.
FSS:	*Cherokee One Four Six One Tango, Shreveport Radio.*
You:	Shreveport, Cherokee Six One Tango. Would you read me SIGMET Delta Four?
FSS:	*Cherokee Six One Tango, SIGMET Delta Four reads as follows: Flight precaution eastern Louisiana, moderate to severe turbulence in clouds seven thousand to fifteen thousand feet MSL. Conditions expected to continue until zero three hundred Zulu.*
You:	Roger, Shreveport. Cherokee Six One Tango.

SIGMETs report severe to extreme turbulence, severe icing, and dust or sand storms that reduce visibility below three miles. In effect, they report all of the more hazardous conditions *except* thunderstorms, tornados, and hail.

Convective SIGMETs focus on these latter phenomena, and, as with AIRMETs and SIGMETs, are broadcast on the voice facilities of Flight Service Stations, as well as being available for the pilot's preflight review. Beginning daily at 0000Z, Convective SIGMETs are numbered consecutively from 01 to 99, with each valid for one hour.

HIWAS (Hazardous In-Flight Weather Advisory Service), a new TWEB-like continuous broadcast service, will soon be available over many navaids to disseminate important weather advisories, such as SIGMETs and AIRMETs.

PIREPs (Pilot Reports) are perhaps the most significant and accurate sources of current flight conditions. Who knows better than the pilot who is there now and can report what is actually happening? Light turbulence that was forecast turns out to be moderate or severe. Unpredicted icing is experienced. Winds that were supposed to be from 240 degrees at 15 suddenly become brutal headwinds or helpful tailwinds at 45 knots. A flock of waterfowl finds its way into the traffic pattern. A sudden wind shear is encountered on takeoff or landing. All of these unexpected events should be immediately reported to the tower, Flight Service, and/or UNICOM (depending on the facility at the particular airport).

As a general rule, submit a PIREP whenever conditions occur that were not forecast or that are actually or potentially hazardous to flight—weather-related or otherwise. Current PIREPs help forecasters, briefers, controllers, and pilots alike. (Flight Watch even appreciates reports of favorable conditions.)

What should be reported, keeping in mind the general rule stated above? Basically, unanticipated icing, clear air turbulence, wind shears, thunderstorms, ceiling or visibility changes (from those forecast), significant wind direction or velocity changes, in-flight or runway obstacles, precipitation, and anything else you encounter that was unexpected and could present a danger to other pilots.

To whom should you make the report? Depending on where you are and with whom you are in communication, the report should go to the nearest Flight Service Station, the Air Route Traffic Control Center (if you're using this facility), Flight Watch, the control tower, or, if at an uncontrolled airport, the local UNICOM. These are the sources that will make use of your report and convey it to others orally and (except for UNICOM operators) over the weather teletype network.

An example of a PIREP communicated in flight:

You: Wichita Radio, Cherokee One Four Six One Tango on one two two point four.

FSS: *Cherokee One Four Six One Tango, Wichita Radio.*

You: Wichita, Cherokee Six One Tango PIREP. Cherokee One Eighty over Abilene at two two three zero Zulu, encountering continuous moderate chop between Topeka and present location. Clear of clouds at eight thousand five hundred.

FSS: *Cherokee Six One Tango, roger. Keep us advised if conditions continue or worsen.*

You: Roger, will do. Cherokee Six One Tango.

In the PIREP, include:

- Type of aircraft
- Location
- Time (UTC)
- Conditions you are reporting
- Whether in or out of clouds
- Altitude
- Duration of conditions you are reporting

If the PIREP concerns icing or turbulence, use the approved definitions as detailed in Chapter 6 of the *Airman's Information Manual*. What to you might seem to be severe turbulence could, by definition, be moderate or perhaps even light. Just be sure that you and the person on the ground are speaking the same language. Misuse of official terms because of unfamiliarity with them can result in inaccurate information to other pilots. They are either lulled into a sense of false security (and we know the dangers of that), or they are warned of conditions that don't actually exist, which could cause flight deviations, increased time en route, and increased fuel consumption.

So you should know what you're talking about when you submit a PIREP. But even if you're not certain about the exact conditions you're encountering, according to official definition, a slightly inaccurate PIREP, when unexpected conditions arise, is better than no PIREP at all. Others will appreciate your concern for their well-being.

EXTENDING THE FLIGHT PLAN

You're over central Kansas, en route to Kansas City on a three-hour flight plan. The winds, however, aren't holding up to the forecast. Either the tailwind is less or the headwinds are greater. Whatever the reason, you're not going to make the three-hour estimate. It looks like the en route time will be closer to a three and a half hours, if not a little longer.

Keeping in mind that the FSS starts asking questions if you haven't closed out your flight plan within 30 minutes of your estimated arrival, you decide to extend the flight plan, based on the new ETA. If you're in the vicinity of Hays, the call would go like this:

You: Wichita Radio, Cherokee One Four Six One Tango, on Hays, one two two point three.

FSS: *Cherokee One Four Six One Tango, Wichita Radio.*

You: Russell, Cherokee One Four Six One Tango over Hays at seven thousand five hundred on VFR flight plan to Kansas City Downtown, with a one five zero zero local ETA. Would like to extend the ETA to one five three zero.

FSS: *Cherokee Six One Tango, roger. Will extend your VFR flight plan to one five three zero local. Russell altimeter three zero zero four.*

You: Three zero zero four. Roger, thank you. Cherokee Six One Tango.

AMENDING THE FLIGHT PLAN IN FLIGHT

The flight is from Memphis to Kansas City, a distance of 410 statute miles. With forecast winds of about 20 knots from 270 to 290, you plan, without hesitation, to make a pit stop at Springfield, Missouri. Along the way, however, you find the winds are much less than anticipated and that you have plenty of fuel to reach Kansas City nonstop. After passing the Dogwood VOR, 35 miles southeast of Springfield, you tune to the Springfield VOR and continue on Victor 159 toward the station.

In checking the Sectional, you note the solid square in the lower right corner of the VOR box, indicating that the station transmits Transcribed Weather Broadcasts (TWEBs). With this service available to you, you monitor the broadcast on the VOR frequency and find that the conditions into Kansas City justify the elimination of the Springfield stop. So you call the Columbia AFSS:

You: Columbia Radio, Cherokee One Four Six One Tango on one two two point five five.

FSS: *Cherokee One Four Six One Tango, Columbia Radio.*

You: Columbia Cherokee One Four Six One Tango is 20 miles southeast on Victor One Five Niner at eight thousand five hundred on VFR flight plan Memphis to Springfield. We have two point five hours of fuel remaining and would like to amend our flight plan and proceed direct to Kansas City Downtown, with an ETA of one four four five local.

FSS: *Cherokee Six One Tango. Understand you have two point five hours of fuel remaining. We will amend your flight plan to show you direct to Kansas City Downtown with a one four four five ETA.*

You: Roger. Thank you for your help. Cherokee Six One Tango.

CHECKING RESTRICTED AREAS OR MILITARY OPERATIONS AREAS (MOAs)

The country is full of areas that are designated as *Prohibited, Restricted, Warning,* or *Alert.* And there are also the *Military Operations Areas.* Suffice it to say that flight into a Prohibited Area is *prohibited;* these are sufficiently few in number and limited in size that they are generally easy to circumnavigate. Many Restricted Areas are also relatively easy to avoid. If flight through such an area, however, is essential, authorization from the controlling agency is required. Entering an active Restricted Area without authority could expose you to all sorts of military hazards, from aerial gunnery to guided missiles. By checking the Special Use Airspace table on the Sectional, you'll find that most Restricted Areas have regular hours of use. But some can be activated by NOTAM on an irregular basis. So you'll need to check with the controlling agency listed in the table. In most cases the controlling agency is the Air Route Traffic Control Center.

Warning Areas are located offshore beyond the three-mile limit, but Alert Areas can be found throughout the country. Normally, the latter define the geography over which there is a high volume of flight training or unusual aerial activity. Transiting these areas is not prohibited, nor is special authorization required. The pilot, however, is urged to exercise extreme vigilance and caution. The responsibility for collision avoidance is his.

Military Operations Areas present the largest geographical obstacle, and to circumvent one might take you miles out of your way. Rather than consume fuel that could otherwise be conserved, the safety of passage through or into an MOA can be determined by contacting any FSS within 100 nautical miles of the area. Perhaps you can do this on the ground during the preflight briefing. Otherwise, call the nearest FSS before entering the MOA.

For example: You're making a trip from Memphis to Jacksonville, Florida. The most direct routing is to Birmingham, Tuskegee, Albany (Georgia), and on to Jacksonville. Just east of Albany, however, lies the extensive Moody MOA— Moody 1, 2A, and 2B. To avoid the MOA would add at least 55 statute miles to the flight—not a major detour, but a needless one in time and fuel, if it can be avoided.

The answer, then, is to contact Macon AFSS to determine the current activity in the area and the extent to which passage through it is feasible. The commonly used word to denote the existence of military operations is "hot."

You:	Macon Radio, Cherokee One Four Six One Tango on one two two point two over Albany VOR Radio.
FSS:	*Cherokee One Four Six One Tango, Macon Radio.*
You:	Macon, Cherokee One Four six One Tango is ten miles northwest on Victor One Five Niner at seven thousand five hundred en route Jacksonville on VFR flight plan. Can you advise if the Moody MOAs are hot?
FSS:	*Cherokee Six One Tango, affirmative. Moody 1A and 2A are both hot at this time. Contact Jacksonville Center for advisories, frequency one three two point five five.*
You:	Roger, will do. Cherokee Six One Tango.

If you've checked the Sectional, you know the floors and ceilings of the activity in the MOA, and because the area is hot, even minimum discretion would tell you that flight between the floors and ceilings should be studiously avoided. Even then, that doesn't say that all is clear. So a call to Center for vectors or advisories is in order.

Obviously, what Center tells you will dictate your actions—to go through the MOA, or take the longer way around. You're not prohibited from entering the MOA, but if there's enough high-speed activity buzzing around in there, good judgment might tell you to follow the airways and add another half hour or so to your journey.

CLOSING OUT THE FLIGHT PLAN

One way to close out a flight plan is by telephone at your destination. The other way is via radio after you're on the ground and parked at the ramp. There can be some exceptions when a close-out in flight is in order, but the best approach is to wait until you've landed and the flight is completed.

Whichever method you choose, just be sure you *do* close it out! Failure to do so sets a lot of wheels in motion to track you down, and this does not sit very well with the powers that be.

If the FSS or an RCO is on the field and you select the radio, the call is simply this:

You: Jackson Radio, Cherokee One Four Six One Tango on one two two point two.

FSS: *Cherokee One Four Six One Tango, Jackson Radio.*

You: Cherokee One Four Six One Tango is on the ground at Memphis International. Please close out my VFR flight plan from Springfield, Missouri, at this time.

FSS: *Cherokee Six One Tango, roger. Closing out your flight plan at three five.*

You: Understand we're closed at three five. Thank you. Cherokee Six One Tango.

OBTAINING SPECIAL VFR: FLIGHT SERVICE REMOTED, NO TOWER

You want to depart from an airport *within a Control Zone*. No tower is on the field, the FSS is physically located elsewhere, and the weather is less than 1,000 feet and three miles visibility. Once clear of the Control Zone, you know that you can maintain legal VFR limits of visibility and cloud separation. How do you obtain a clearance under these conditions?

Let's take a specific example. Gage, Oklahoma's Gage-Shattuck Airport has only UNICOM—no tower, no Flight Service Station. There is, however, a Remote Communications Outlet through which you can contact the McAlester FSS more than 200 miles to the southeast. The remoted facility is identified on the Sectional (FIG. 4-6).

The weather being what it is, you can't legally take off in the Control Zone without clearance through Flight Service. Hence, this call:

You: McAlester Radio, Cherokee One Four Six One Tango, on Gage, one two two point five five.

FSS: *Cherokee One Four Six One Tango, McAlester Radio.*

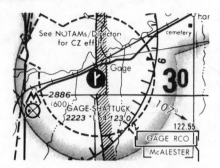

Fig. 4-6. *The Gage-Shattuck Airport has no tower, the FSS is in McAlester, and the airport lies within a Control Zone.*

You: McAlester, Cherokee One Four Six One Tango requests Special VFR out of the Gage Control Zone, west departure.

FSS: *Cherokee Six One Tango, stand by.* [FSS now contacts the appropriate control agency.]

FSS: *Cherokee Six One Tango, ATC clears Cherokee One Four Six One Tango to exit the Gage Control Zone to the west. Maintain Special VFR conditions at or below three thousand feet while in the Control Zone. Report leaving the Control Zone.*

You: Roger, understand Cherokee Six One Tango is cleared out of the Gage Control Zone to the west, to maintain Special VFR at or below three thousand in the Zone. Will report leaving the Zone.

FSS: *Cherokee Six One Tango, readback is correct.*

You: Roger. Cherokee Six One Tango.

Remember that in a non-radar area, only one aircraft at a time is allowed in the Control Zone when conditions are below normal VFR limits. Thus for safety first and courtesy to others second, it is imperative to advise Flight Service when you are clear of the Zone. (Once again, as we said in the UNICOM section, we'll get into complete definitions/descriptions of Control Zones, controlled airports, Airport Traffic Areas, and the like in Chapter 8.) When clear of the Zone, call McAlester Radio via Gage, as above, and report:

You: Cherokee One Four Six One Tango is clear of the Gage Control Zone to the west.

FSS: *Cherokee Six One Tango, roger. Gage altimeter two niner zero eight.*

You: Two niner zero eight. Thank you for your help. Cherokee Six One Tango.

Now let's reverse the process. You want to enter the Control Zone and land at Gage. The weather was forecast to be below VFR limits, and a call to Gage

UNICOM verifies that fact. Once again, Flight Service comes into the picture:

> **You:** McAlester Radio, Cherokee One Four Six One Tango, on Gage, one two two point five five.
>
> *FSS:* *Cherokee One Four Six One Tango, McAlester Radio.*
>
> **You:** Cherokee One Four Six One Tango is 15 miles west of Gage and requests Special VFR to enter the Control Zone for landing at Gage.
>
> *FSS:* *Cherokee Six One Tango, we have one other aircraft in the Control Zone for landing. Stand by and remain clear of the Zone until further advised.*

Now, find an area well outside the Zone and hold until the FSS authorizes you to proceed:

> *FSS:* *Cherokee Six One Tango, McAlester Radio.*
>
> **You:** McAlester Radio, Cherokee One Four Six One Tango, go ahead.
>
> *FSS:* *Cherokee Six One Tango, ATC clears Cherokee One Four Six One Tango to enter the Gage Control Zone. Maintain Special VFR conditions at or below three thousand feet while in the Control Zone. Report when down and clear at Gage.*
>
> **You:** Understand, Cherokee Six One Tango cleared to enter the Gage Control Zone at or below three thousand on Special VFR. Will report on the ground at Gage.
>
> *FSS:* *Cherokee Six One Tango, readback correct.*
>
> **You:** Roger. Cherokee Six One Tango.

You're on the ground and clear of the active at Gage. Now call Flight Service over the RCO as quickly as possible so that another aircraft can be cleared into or out of the Zone:

> **You:** McAlester Radio, Cherokee One Four Six One Tango is down and clear at Gage.
>
> *FSS:* *Understand Cherokee Six One Tango is down and clear at Gage.*
>
> **You:** That is affirmative—and thank you for your help. Cherokee Six One Tango.

OBTAINING SPECIAL VFR:
FLIGHT SERVICE ON THE AIRPORT, NO TOWER

Emporia, Kansas, (about 80 miles southwest of Kansas City) has a Control Zone, no tower, and until recently the FSS was located on the airport. (Recall the excerpt from the Sectional, FIG. 3-2?) Before departing Emporia, you check the weather by phone or in person, file a flight plan if the flight so warrants, and advise Flight Service that you'll be requesting a Special VFR because of the local weather conditions. In all likelihood, Flight Service, through ATC, won't be able to give the final clearance until you are ready to take off, because the minimums might drop below the Special VFR limits between briefing and take-off times, your estimated departure might be delayed, or other aircraft could be entering or leaving the Control Zone when you are ready to go.

Once you are ready, however, the radio contacts are the same as those illustrated in the Gage situation. Whether the FSS is on the airport (which is becoming increasingly unlikely) or remoted makes no difference. The procedures for departing and entering the Zone are identical.

OBTAINING AIRPORT ADVISORIES: FLIGHT SERVICE ON THE AIRPORT, NO TOWER

When an operating Flight Service Station is located *on* the airport and there's no operating tower, the FSS provides the same service that UNICOM offers at fields with no tower or FSS, but more detailed and more accurate. Using Emporia again, the on-field existence of the FSS is indicated by the heavy lined box on the Sectional (FIG. 4-7).

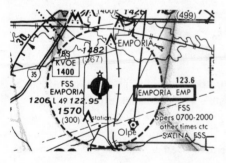
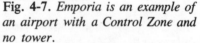

Fig. 4-7. *Emporia is an example of an airport with a Control Zone and no tower.*

If you're landing at an airport with an FSS and want an airport advisory, what frequency do you use? The answer is the same at all airports: 123.6. This is the standard frequency for the AAS, or Airport Advisory Service. It is *not* the frequency for filing or closing out flight plans or for obtaining en route weather information. Going into a field which has *both* Flight Service and UNICOM, use the latter for phone messages, taxis, and the like, and the former for the advisory service. Accordingly, and on 123.6, the communication goes like this:

You:	Emporia Radio, Cherokee One Four Six One Tango on one two three point six.
FSS:	*Cherokee One Four Six One Tango, Emporia Radio.*
You:	Emporia, Cherokee Six One Tango is ten miles west of Emporia VOR at three thousand. Request airport advisory.
FSS:	*Cherokee Six One Tango, favored runway is One Eight, wind one niner zero at one two, altimeter two niner eight seven. No reported traffic.*
You:	Roger. Thank you. Cherokee Six One Tango.

KEEPING LOCAL TRAFFIC INFORMED

Remember that Flight Service is in no way a traffic control agency at Emporia or anywhere else. So once you enter the pattern, stay on 123.6 but address your communiques to "Emporia *Traffic*," not "Emporia *Radio*."

From this point on, the calls are the same as at a MULTICOM or UNICOM field—entering the pattern on downwind, turning base, turning final, and when on the ground and clear of the runway.

As with UNICOM, don't expect any response from the FSS when you make these transmissions. We've noticed on several occasions, however, particularly at the smaller or less busy airports, that the specialist does acknowledge even routine position reports. This despite the pilot's use of "Traffic" rather than "Radio." So if the FSS does respond to or acknowledge blind transmissions, don't be surprised. On the other hand, don't be surprised if silence reigns. That's the way it should be.

Remember that there's no *requirement* for you to contact Flight Service for an advisory. Not doing so is a little foolish, though. Fulfill *your* responsibility, but keep your eyes open and your head turning to spot any possible Silent Sams.

CLOSING OUT A VFR FLIGHT PLAN:
FLIGHT SERVICE ON THE AIRPORT, NO TOWER

At the Emporias of the nation, when you land and want to cancel your flight plan, you can do so by radio, if that's your choice. But just be sure to switch from 123.6 to another published frequency—in the case of Emporia, 122.2:

You:	Emporia Radio, Cherokee One Four Six One Tango on one two two point two.
FSS:	*Cherokee One Four Six One Tango, Emporia Radio.*
You:	Cherokee One Four Six One Tango is on the ground at Emporia. Would you close my flight plan from Wichita at this time?

> *FSS:* *Cherokee Six One Tango, roger. Will close out your flight plan at two seven.*
>
> You: Roger. Thank you. Cherokee Six One Tango.

TAXIING OUT AND BACK-TAXIING: FLIGHT SERVICE ON THE AIRPORT, NO TOWER

As we said in the MULTICOM and UNICOM chapters, if you're going to back-taxi, complete the entire pre-takeoff check before venturing onto the runway—*unless* an area exists at the end of the runway for that purpose. It's simply discourteous to others and disruptive to traffic to park on the approach end while you go through the checklist. More than one landing pilot has been forced to go around because some self-centered idiot was sitting there on the runway, oblivious to the existence of all others. If there is no runup area, make the check on the ramp or taxiway, back-taxi with some speed and power, do a 180, and get going.

But to begin at the beginning: you're on the ramp, engine started, and you want an airport advisory. Tune to 123.6 and call Flight Service:

> You: Emporia Radio, Cherokee One Four Six One Tango on the ramp, ready to taxi for takeoff. VFR northbound. Request airport advisory.
>
> *FSS:* *Cherokee Six One Tango, favored runway is One Eight, wind one niner zero at four. Altimeter three zero zero zero. Mooney reported ten west for landing.*
>
> You: Roger, will back-taxi on One Eight after runup. Cherokee Six One Tango.

You've completed the pre-takeoff check and are ready to taxi:

> Emporia *Traffic*, Cherokee Six One Tango back-taxiing on Runway One Eight, Emporia.

You're ready to go:

> Emporia Traffic, Cherokee Six One Tango departing on Runway One Eight, northwest departure, Emporia.

CTAFs: PART/FULL-TIME FSS, PART/FULL-TIME TOWER

Budget-cutting in recent years has brought about a profusion of part-time towers and part-time FSSs. As a general rule, when a tower exists on a field, even if it is closed, use the tower frequency for traffic advisories. If there is no tower but an FSS exists on the field, use 123.6 for self-announce advisories even if the FSS is closed. Consult the chart below, but remember to check the *A/FD* for exceptions:

Tower Status	FSS Status	Frequency
Open Tower on Field		Tower-controlled on Tower Frequency
No Tower on Field	Open FSS on Field	AAS on 123.6
No Tower On Field	Closed FSS on Field	Self-announce on 123.6 Advisories from UNICOM (if available)
No Tower on Field	No FSS on Field	UNICOM (if available) or MULTICOM (check *A/FD*)
Closed Tower on Field	Open FSS on Field	AAS on Tower Frequency
Closed Tower on Field	Closed (or No) FSS on Field	Self-announce on Tower Frequency

Check the current edition of the *A/FD* for any exceptions to this chart.

And one more summary to help keep things straight when there is no tower, but a Flight Service Station as well as UNICOM is on the field.

Purpose of Call	Frequency	Radio Call
For airport advisories (when FSS is open)	123.6	"_____ Radio"
Traffic position reports (at all times)	123.6	"_____ Traffic"
Flight plan filing/closing	Published FSS frequency or 122.2	"_____ Radio"
To order taxis, relay messages, request mechanics, etc. (and airport advisories when FSS is closed)	Published UNICOM freq. (usually 122.7, 122.8, or 123.0)	"_____ UNICOM"

DIRECTION FINDING (DF) ORIENTATION

You're lost, or unsure of your position. As a rule, if you have even one navcom on board, you should be able to reorient yourself with little difficulty. It's just a matter of tuning to one VOR, centering the needle, drawing the radial from that station on your Sectional, and then doing the same thing with a second VOR. Your position is where the lines intersect. Sometimes, however, we can be in an area or at an altitude where one or more VORs are out of receiving range, or, indeed, the navcoms could go kaput. Now we need help. Here is where a Flight Service Station can provide "Emergency Assistance Service," help us to get oriented, and, if necessary, provide us with a DF Steer to the nearest airport.

Briefly summarized, depending on the ground equipment available, the FSS, in coordination with the pilot, offers five types of DF orientation:

1. Almost immediate location determination with the FSS using two geographically separated DF receivers. One receiver is usually in the FSS facility, the other at a remote location;

2. Orientation with one DF receiver and the pilot tuning to a VOR, when that VOR is within range of the aircraft's reception, to obtain the radial from the VOR, and thus a "cross-fix" identifying the aircraft's position;

3. DF-T/D (Time and Distance), when the FSS uses only one DF and no other means of obtaining a cross-fix is available;

4. DF-ADF, using one DF receiver, with the pilot tuning to a nondirectional beacon (NDB) within the aircraft's reception range—if the aircraft is equipped with Automatic Direction Finding (ADF) equipment. This produces a cross-fix, similar to the DF-VOR orientation in 2., above;

5. VOR-to-VOR, when no DF equipment is available (or is beyond the aircrafts' range), the aircraft is within range of two VORs, and has at least one operating VOR receiver on board. This type of orientation the pilot could do by himself, if he had a current Sectional on board, but inexperience, increasing anxiety, or other factors might require the help of the FSS specialist.

Rather than trying to illustrate what would take place in each of the five orientation possibilities, let's focus just on the first, where there is a functioning VOR on board and the FSS uses two DF receivers—one at the station and one remote. The situation, then, is this:

You're heading west in western Iowa, where the land is flat and easy-to-spot landmarks are infrequent. You know generally that you're somewhere southwest of Fort Dodge because, with the needle centered, the VOR head shows a 45 degree heading to the Fort Dodge VOR. All other VORs are out of range, however, a study of the terrain reveals nothing identifiable, and your fuel is rapidly diminishing. Time is critical, so you call the Fort Dodge (FOD) Automated Flight Service Station for emergency assistance.

First, what is the general sequence of information the AFSS specialist will request, and what instructions will he likely issue? There may well be variances, but the following is in line with current AFSS procedures:

1. The *minimum* information the AFSS will want is aircraft identification, aircraft type, if transponder-equipped, pilot intentions, and nature of the emergency. Additionally, the specialist may ask the weather at your altitude, the hours of fuel remaining, and possibly (or probably) to squawk the "mayday" 7700 transponder code. If the last is the case, the specialist will then coordinate the orientation process with the appropriate Center or airport Approach Control.

2. Other likely instructions would tell you to maintain VFR flight, advise of any necessary altitude changes, and advise if AFSS instructions would cause you to violate VFR regulations. The specialist will also give you the current altimeter setting, ask you to set the directional gyro heading indicator to the magnetic compass heading, and then request you to report current heading and altitude.

3. Instructions will be included for keying the microphone by depressing the mike button for 10 seconds, saying nothing during that period, and concluding the transmission with your aircraft identification.

4. And there will be instructions during the actual orientation process. If the AFSS is using two DF receivers, your position will be determined very rapidly. More extensive directions are probable if the AFSS is using only one DF and wants you to tune to a VOR for a cross-fix. The same applies when the orientation involves one DF and an NDB, a DF/time-and-distance procedure, or the VOR-to-VOR method.

5. Finally, if necessary, the AFSS will give you headings, or a "DF Steer," to the nearest airport or the airport of your choice, depending upon the situation.

You:	Fort Dodge Radio, Cherokee One Four Six One Tango on one two two point two.
FOD:	*Cherokee One Four Six One Tango, Fort Dodge Radio.*
You:	Fort Dodge, Cherokee Six One Tango is somewhere southwest of you at six thousand five hundred. Can only pick up the Fort Dodge VOR and have three zero minutes of fuel remaining. Request a DF steer to the nearest airport.
FOD:	*Roger, Six One Tango. Are you transponder equipped?*
You:	Affirmative, Fort Dodge.
FOD:	*Six One Tango, squawk seven seven zero zero.*
You:	Roger, seven seven zero zero. Six One Tango.
FOD:	*Six One Tango, what are the weather conditions at your altitude?*
You:	Scattered cumulus, Fort Dodge.

FOD: *Roger, Six One Tango. Remain VFR at all times, and advise if you have to change heading or altitude to remain VFR. Fort Dodge altimeter is three zero zero five.*

You: Roger, three zero zero five.

FOD: *Six One Tango, maintain straight and level flight and reset your heading indicator to agree with your magnetic compass. After you have done this, say your heading and altitude.*

You: Fort Dodge, Six One Tango heading is two seven zero degrees, level at six thousand five hundred.

FOD: *Roger, Six One Tango. Now depress your mike button for 10 seconds, saying nothing, and then follow with your aircraft identification.*

You: Roger. (Depress the mike button 10 seconds,) Cherokee One Four Six One Tango.

    (Note: During the 10 seconds, the AFSS specialist is watching the strobes on the two DF receivers, as the strobes search out your aircraft based on your radio transmission. Where the strobes intersect identifies your position. The specialist then plots that position on a Sectional chart.)

FOD: *Cherokee Six One Tango, your position is one five miles northeast of the Denison Airport. What are your intentions?*

You: We need to land Denison as soon as possible for fuel. Six One Tango.

FOD: *Roger, Six One Tango. For a heading to the Denison Airport, turn left to two two zero degrees. Report when on that heading.*

You: Roger, left to two two zero.

You: Fort Dodge, Six One Tango on the three two zero heading.

FOD: *Roger, Six One Tango. Do you see any prominent landmarks?*

You: Affirmative. We see a water tower and a town in the distance at twelve o'clock.

FOD: *Roger, Six One Tango. That should be Denison. Are you familiar with the airport?*

You: Negative. Please advise.

FOD: *Roger, Six One Tango, the Denison Airport is three miles southwest of the town, elevation one thousand two hundred seventy three feet. The one runway is northwest-southeast. Report when the airport is in sight.*

You: Roger, will do.

You: Fort Dodge, Six One Tango has the airport.

FOD: *Roger, Six One Tango. The Denison UNICOM CATAF is one two two point eight and Denison altimeter is three zero zero zero.*

You: One two two point eight and three zero zero zero.

FOD: *Roger, Six One Tango, do you require any further assistance?*

You: Negative, Fort Dodge. Six One Tango is in good shape now. Thank you for your help.

FOD: *Roger, Six One Tango. DF orientation services is terminated. Good day.*

5

Automatic Terminal Information Service (ATIS)

Before continuing, it would be appropriate to mention that there is a certain logic to this book. It attempts to take you through the normal sequence of radio communications that you would experience in a typical flight, whether that flight is limited to the immediate vicinity of an airport or is a cross-country. The only difference between the two is the facilities that come into play on a cross-country, such as Approach Control and Center, which are usually not involved in a purely local excursion.

The book began with a discussion of MULTICOM and UNICOM, the simplest of all communication procedures, merely because they *are* the simplest and are practiced primarily at uncontrolled airports. Then came a description of Flight Service, how to contact it, how to use its services, and its role of providing essential information to those who venture beyond the local pea patch. In most cases, Flight Service represents the initial contact with the National Airspace System. It's where the weather is determined and the flight plan is filed. It's typically the first step in the sequence of communications between pilot and ground personnel.

For a complete explanation of Flight Service, however, we did include the various in-flight communications that take place. In that sense, the normal sequence of events from takeoff to landing was violated. Now each facility will be discussed sequentially (with possibly a few exceptions) beginning with ATIS, then Ground Control, Tower, Approach/Departure Control, and finally Center.

WHAT IS ATIS?

The Automatic Terminal Information Service, with its neat little acronym of ATIS, is hardly a stranger to most pilots—especially those acquainted with the busier airports. As a source of important information, it is a vital communications tool for both the departing and arriving airman.

For those not very familiar with it, ATIS is a continuous recorded summary, not more than one hour old, of local weather, winds, and the runway(s) and instrument approach(es) currently in use. Its purpose is to provide the pilot with the basic airport data on a frequency that does not interfere with live radio communications. By so doing, the controllers can concentrate on controlling traffic without having their attention diverted by transmitting weather, winds, altimeter, runways in use, and the like.

Recorded in—and transmitted from—the tower, ATIS information is normally updated every hour, and is identified by the phonetic alphabet—Information Alpha, Bravo, Charlie, etc. Should conditions change in any material way before the next scheduled update, a revision is issued accordingly and given the next phonetic designation.

DETERMINING THE APPROPRIATE FREQUENCY

If ATIS is available at a given airport, its frequency is indicated on the Sectional and in the *Airport/Facility Directory*.

To illustrate from the Sectional, take Smith Reynolds Airport in Winston-Salem, North Carolina, as an example (FIG. 5-1). In this case, the ATIS is clearly identified as 121.3. If no frequency is stated, the service is not provided, and winds, altimeter, and runway are obtained from the tower. At some locations, ATIS is transmitted over the voice facility of a nearby VOR or VORTAC. At some of the larger airports, where different runways are used for takeoffs and landings, two ATIS frequencies are used—one for departures, the other for arrivals (FIG. 5-2). The dual transmissions make for shorter messages and reduce the likelihood of pilot misinterpretation.

Fig. 5-1. *How the Sectional identifies the ATIS (Automatic Terminal Information Service) frequency. In the case of Smith Reynolds, it is 121.3.*

§ **NEWARK INTL** (EWR) 2.6 S UTC–5(–4DT) 40°41'35"N 74°10'07"W   **NEW YORK**
   18 B S4 FUEL 100LL, JET A OX 3 LRA CFR Index E   H-3D, L-24H, 25B, 28G
RWY 04R-22L: H9300X150 (ASPH-GRVD)   D-191, DT-358, DDT-823   HIRL CL   **IAP**
  RWY 04R: ALSF2. TDZ. Thld dspicd 1190'. Pole.
  RWY 22L: SSALR. TDZ. VASI(V4L) Thld dspicd 1090'. Fence.
RWY 04L-22R: H8200X150 (ASPH-CONC-GRVD)   D-191, DT-358, DDT-823   HIRL CL
  RWY 04L: SSALR. TDZ. Thld dspicd 740'.
  RWY 22R: REIL. VASI(V6L)—Upper GA 3.25°TCH 85', Lower GA 3.0°TCH 52'. Thld dspicd 440'. Antenna.
RWY 11-29: H6800X150 (ASPH-CONC-GRVD)   S-191, D-358, DT-568, DDT-823   HIRL
  RWY 11: VASI(V4L)—GA 3.08°TCH 53'. Building.
  RWY 29: REIL. VASI(V6L)—Upper GA 3.25°TCH 85', Lower GA 3.0°TCH 52'. Thld dspicd 298'. Highway.
AIRPORT REMARKS: Attended continuously. Landing fee. Flocks of birds on and in vicinity of airport. Control tower visual
  blind spot on inner and outer taxiways east of taxiway 'A'; heliport operations on inner and outer taxiways adjacent
  'RA' intersection 1200-0300Z‡. Flight Notification Service (ADCUS) available. NOTE: See Special No-
  tices—Operations Reservations for High Density Traffic Airports.
WEATHER DATA SOURCES: LLWAS.
COMMUNICATIONS: ATIS 115.7 (ARR) 132.45 (DEP) 201-624-6463.   UNICOM 122.95
  TETERBORO FSS (TEB) Toll free call, dial 1-800-932-0835. NOTAM FILE EWR
Ⓡ **NEW YORK APP CON** 127.6 (270°-089°) 128.55 (090°-269°) 126.7   Ⓡ**NEW YORK DEP CON** 135.35 119.2
  TOWER 118.3 134.05   **GND CON** 121.8 126.15   **CLNC DEL** 118.85   **PRE-TAXI CLNC** 118.85
  TCA Group I: See VFR Terminal Area chart.
RADIO AIDS TO NAVIGATION: NOTAM FILE TEB.
  TETERBORO (T) VOR/DME 108.4   TEB   Chan 21   40°50'55"N 74°03'46"W   216°10 NM to fld. 10/11W.
    Unmonitored indefinitely.
  PROGRESS NDB (H-SAB) 379   ▪ GKQ   40°40'54"N 74°11'30"W   058°1.0 NM to fld.
  CHESA NDB (LOM) 241   EW   40°35'37"N 74°13'49"W   039° 5.8 NM to fld.
  LIZAH NDB (LOM) 204   EZ   40°36'26"N 74°13'06"W   039° 4.9 NM to fld.
  ILS 108.7 I-LSQ Rwy 22L. Localizer unusable beyond 015°on left side of course.
  ILS 108.7 I-EWR Rwy 04L LOM CHESA NDB Localizer unusable beyond 020°on left side at all altitudes.
  ILS 108.7 I-EZA Rwy 04R LOM LIZAH NDB

Fig. 5-2. *Newark International not only has separate ATISs for arrivals and departures, but you can also telephone the number listed to listen to one of the recordings.*

INFORMATION PROVIDED BY ATIS

Assuming no unusual or potentially hazardous conditions, the typical ATIS includes the following data:

☒ Location

☒ Information code (phonetic alphabet)

☒ Time (UTC, "coordinated universal time," stated as "Zulu")

☒ Sky conditions (often omitted if ceiling is higher than 5000 feet, or is stated as "better than 5000")

☒ Visibility (often omitted if visibility is greater than five miles, or is stated as "better than five miles")

☒ Temperature and dewpoint

☒ Wind direction (magnetic) and velocity

☒ Altimeter setting

☒ Instrument approach in use

✈ Current runway in use

✈ NOTAMs (if any)

✈ Information code repeated (phonetic alphabet)

A typical example:

> Jackson Information Delta. One six four five Zulu weather. Four thousand five hundred broken, visibility five. Temperature four five, dewpoint three eight. Wind two zero zero at one six, gusting to two two. Altimeter three zero zero niner. ILS Runway One Five in use, land and depart Runway One Five. Advise you have Delta.

Depending on the facility, some ATIS tapes will include departure and approach frequencies and the clearance delivery frequency. In the event of adverse weather, ATIS broadcasts will include AIRMETs and SIGMETs, such as "SIGMET Charlie Two is now in effect." To determine what Charlie Two is all about, you have to contact Flight Service. ATIS does not provide all of the details.

To illustrate some of the additional information that could be included in an ATIS broadcast, let's take this example:

> Jackson Information Foxtrot. One three three zero Zulu special observation. Five hundred scattered, estimated ceiling one thousand overcast, visibility five, light rain. Temperature seven zero, dewpoint six two. Wind three two zero at seven. Altimeter two niner niner zero. VOR Delta approach in use. Left traffic. Land and depart Runway Three Five. SIGMETs Charlie Two Niner, Charlie Three Zero, and Charlie Three One now in effect. Numerous trenches along Runway Three Five. Advise you have Foxtrot.

WHEN TO TUNE TO ATIS

Departure: The engine is started; radios and beacons are on. Before any further action, tune to the ATIS frequency and *listen.* As it is a continuous transmission, listen *twice*—especially if the details of the first transmission weren't clear. Set the altimeter, and remember the information code.

Arrival: Twenty or 25 miles out, tune to the ATIS frequency and again listen—perhaps a couple of times so that you have the details firmly in mind. Presumably you're not too busy that distance from the field, so you can give attention to what's being transmitted.

COMMUNICATING THE FACT
THAT YOU HAVE THE ATIS

Tell those who need to know that you have the current ATIS that you do indeed have it. This includes Ground Control for departure, and Approach Control or the tower when landing. Examples follow:

To Ground Control (when you're ready to taxi):

You: Jackson Ground, Cherokee One Four Six One Tango at Avitat with Information Foxtrot. Ready to taxi. VFR New Orleans.

GC: *Cherokee One Four Six One Tango, taxi to Runway Three Five.*

You: Roger. Taxi to Three Five. Cherokee One Four Six One Tango.

To the Tower (You're inbound and over an identifiable checkpoint 15 miles or so from the field. You have not used Approach Control, or, if you have, you were notified that "radar service is terminated." In other words, this is your first contact with the tower):

Memphis Tower, Cherokee One Four Six One Tango over Collierville, level at three thousand five hundred, squawking one two zero zero with Information India.

To Approach Control (Center has handed you off to Approach):

Kansas City Approach, Cherokee One Four Six One Tango is with you, level at five thousand five hundred, with Information Kilo.

If you fail to indicate that you have the ATIS information, you may well be questioned: "Cherokee Six One Tango, do you have ATIS?" Or the tower or Approach may offer the basic data necessary for your entry into the pattern and for landing. That's a nice gesture, but it consumes air time.

USING THE PHONETIC DESIGNATION

If you've received the ATIS, make it clear to the controller that you have Information Alpha, Bravo, Charlie, or whatever. That informs the controller that you are in possession of the most current and complete information.

Perhaps it's sort of classy to come out with "have the numbers," but it's misleading. It means to the controller that you have received the wind, altimeter setting, and runway only—*not* the full ATIS, which may include SIGMETs, NOTAMs, ceilings, visibility, and so on. Hearing "have the numbers" may cause the controller to question you further as to the extent of your information or require him to relay significant information that is included in the latest ATIS.

Another thing: merely reporting that you "have the ATIS" is not enough. Maybe a special update was just issued. Maybe you had Information Papa 20 minutes ago, but Information Quebec has come out in the meantime. And maybe Quebec designates Runway 18 as the active runway, whereas Papa had it as Runway 26. Any number of changes might have taken place between Papa and Quebec.

To eliminate confusion and reduce air-time talk, be brief but be complete. Use the phonetic alphabet to identify the ATIS you have received. It takes no longer to be specific—and it's proper procedure. As the *AIM* says, " 'Have the numbers' . . . does not indicate receipt of the ATIS broadcast and should never be used for this purpose."

WHEN UNABLE TO REACH THE ATIS

If, for any reason, you're unable to receive the ATIS or the transmission is garbled, tell Ground Control before taxiing out, or Tower or Approach Control if you're coming in. Just conclude the call with "negative ATIS" or "ATIS garbled." The facility you're contacting will then give you the pertinent data so that you can proceed accordingly.

CONCLUSION

ATIS broadcasts exist to: (1) provide you with the departure or arrival information you need [at worst, the information is only 59 minutes old, and is thus a reasonably accurate depiction of existing conditions], and (2) reduce unnecessary communications, thereby permitting pilots and controllers to concentrate on their primary responsibilities—the safe operation of the aircraft and the safe direction of air traffic.

Use ATIS wherever it is available. You don't have to talk back to it. All it requires is the willingness to listen and the ability to remember. If you can't do either of those you shouldn't be in the air in the first place.

6

Ground Control

Most pilots who inhabit the busier airports are well acquainted with Ground Control and the function it plays in assuring the safe operation of earthbound traffic. However, the inexperienced airman, venturing into one of those airports for the first time, needs to understand the responsibilities of Ground Control and the communications with the controller. And so do some of the more experienced aviators. Let's just say that we've heard a lot of hesitant radio calls, incomplete calls, unprepared calls, and excessive verbal garbage—all of which reflects poorly on those who should know better.

What follows should be of some value to *all* general aviation pilots, regardless of their levels of experience.

WHAT DOES GROUND CONTROL DO?

Ground Control is the "invisible policeman" ensconced in the tower who is responsible for the operation of aircraft *and all other vehicles* that are utilizing taxiways and runways *other* than the active runway. It controls the movement of those vehicles through radio communications. It is *the* authority. When the controller says "go," you go; when he says "stop," you stop; when he says "give way," you give way, and so on. If, for any reason, safety or otherwise, you can't comply with his directions, you have the responsibility to advise the controller accordingly. Adherence to his instructions is expected.

At almost every airport, however, there are areas not under the control of Ground, for example, the expanses of concrete used for tiedown purposes, fueling, moving aircraft to and from hangars, compass swinging, and the like. As long as those areas don't infringe on runways or taxiways, you can taxi your aircraft around there all day. Nobody will say a word. Otherwise, you must have your intended movements cleared by Ground Control.

FINDING THE GROUND CONTROL FREQUENCY

The Ground Control frequency will not be found on any Sectional. Where the service exists (at all tower-controlled fields), you can locate the frequency in the *Airport/Facility Directory*, as FIG. 6-1 illustrates.

NEWBURGH

§ **STEWART INTL** (SWF) 3.5 NW UTC−5(−4DT) 41°30′14″N 74°06′19″W **NEW YORK**
    491 B S4 FUEL 100LL, JET A TPA—1500(1009) LRA CFR Index A **H-3D, L-25B, 28H**
    RWY 09-27: H11818X150 (ASPH-GRVD) S-85, D-175, DT-350, DDT-775 HIRL CL **IAP**
      RWY 09: ALSF2. TDZ. VASI(V4L)—GA 3.0°TCH 55′. Thld dsplcd 2000′. Fence.
      RWY 27: REIL. VASI(V6L)—GA 3.0°TCH 50′. Thld dsplcd 2000′. Tree.
    RWY 16-34: H6006X150 (ASPH-CONC) S-36, D-75, DT-140 MIRL
      RWY 16: Trees.   RWY 34: VASI(V4L)—GA 3.0°TCH 59′. Trees.
    AIRPORT REMARKS: Attended continuously. CAUTION—Dutchess County arpt 3.3 NM SW. of KINGSTON Vortac; do not
      mistake for Stewart arpt. Fee for acft over 12,500 lbs. Avoid Orange Co. Arpt, lctd 7 NM WNW during VFR apchs.
      Rgt tfc on rwys 16 and 27 may be used for noise abatement. Rwy 09 ALSF2 with SSALR configuration for CAT 1
      operations. B747 acft req vehicle escort for all taxi operations in main ramp area. Control zone effective
      0500-0459Z‡ Tue-Fri, 0500-0400Z‡ Sat, 1200-0400Z‡Sun, 1200-0500Z‡ Mon. Flight Notification Service (AD-
      CUS) available.
    COMMUNICATIONS: UNICOM 122.95
      POUGHKEEPSIE FSS (POU) Toll free call, dial 1-800-942-8220. NOTAM FILE SWF
      BOSTON CENTER APP/DEP CON 133.1
      TOWER 121.0 NFCT GND CON 121.9 ◄
    RADIO AIDS TO NAVIGATION: NOTAM FILE SWF.
      KINGSTON (L) ABVORTAC 117.6 ■ IGN Chan 123 41°39′55″N 73°49′22″W 244°15.3 NM to fld.
      580/12W. General outlook only 0400-1200Z‡. NOTAM FILE POU.
      NEELY NDB (MHW/LOM) 335 SW 41°29′09″N 74°13′42″W 090°4.8 NM to fld.
      ILS 110.1 I-SWF Rwy 09 LOM NEELY NDB. BC unusable. OM unmonitored.

Fig. 6-1. *Locating the Ground Control frequency in the Airport/Facility Directory.*

One fact you can *almost* always count on: the frequency will be 121.6, .7, .8, or .9. The most common is 121.9, but the other frequencies come into play when two airports are in close physical proximity and using the same frequency would cause confusion.

There are a few exceptions to the general rule of 121.6, 121.7, etc. as at Detroit Metro. (FIG. 6-2). Miami International uses one frequency for one set of runways and another frequency for the other runways. Memphis has one frequency for general aviation, while another is reserved for air carriers. To be sure of your Ground frequency, consult the latest edition of the *A/FD*.

WHEN TO CONTACT GROUND CONTROL

The following are typical situations in which you would contact Ground Control. Some contacts are required, and others are optional, as noted.

```
§   DETROIT METROPOLITAN WAYNE CO    (DTW)   15 S   UTC–5(–4DT)   42°12'55"N 83°20'55"W    DETROIT
      639    B    S4    FUEL 80, 100LL, JET A, A1 +    OX 1, 3    AOE    CFR Index E           H-3C, L-23C, A
      RWY 03L-21R: H10501X200 (CONC-GRVD)   S-100, D-185, DT-350   HIRL CL                    IAP
         RWY 03L: ALSF2. TDZ. Tree.              RWY 21R: MALSR. Railroad.
      RWY 03R-21L: H10000X150 (CONC-GRVD)   S-100, D-200, DT-350, DDT-750    HIRL CL
         RWY 03R: ALSF2. TDZ.                   RWY 21L: MALSR.
      RWY 09-27: H8700X200 (ASPH–CONC–GRVD)   S-100, D-185, DT-350   HIRL
         RWY 09: REIL VASI(V4R)—GA 3.0° TCH 46'.           RWY 27: MALSR. Tree.
      RWY 03C-21C: H8500X200 (ASPH–CONC–GRVD)   S-100, D-185, DT-350   HIRL
         RWY 03C: REIL. VASI(V4L)—GA 3.0° TCH 60'. Tree.   RWY 21C: REIL. VASI(V4L)—GA 3.0° TCH 59.4'.
      AIRPORT REMARKS: Attended continuously. Landing fee. Rwy 03R ALSF2 required when RVR/visibility is 6000/1 mile or
         less. SSALR ops when RVR/visibility is 6000/1 mile. Flight Notification Service (ADCUS) available.
      WEATHER DATA SOURCES: LLWAS.
      COMMUNICATIONS: ATIS 124.55 (313) 942–9350     UNICOM 122.95
         LANSING FSS (LAN) Toll free call, dial 1–800–322–5552. NOTAM FILE DTW
         CARLETON RCO 122.1R, 115.7T (LANSING FSS)
      ®  DETROIT APP CON 124.05, 124.25 (211°-029°) 125.15 (030°-210°)
         METRO TOWER 135.0 (WEST) 118.4 (EAST)  GND CON 121.8 (WEST) 119.45 (EAST)    CLNC DEL  120.65
         PRE TAXI CLNC 120.65
      ®  DETROIT DEP CON 120.15 (030°-210°) 8.95 (211°-029°)
         TCA Group II: See VFR Terminal Area chart.
      RADIO AIDS TO NAVIGATION: NOTAM FILE LAN
         CARLETON (H) VORTAC 115.7    CRL    Chan 104    42°02'53"N 83°27'28"W    033° 10.7 NM to fld. 630/3W.
         WILLOW RUN (T) VORW/DME 110.0    YIP    Chan 37    42°14'13"N 83°31'31"W    102° 7.3 NM to fld. 707/3W.
         REVUP NDB (LOM) 388    DT    42°07'12"N 83°25'55"W    037° 6.4 NM to fld.
         SPENC NDB (LOM) 223    DM    42°13'12"N 83°12'13"W    273° 5.7 NM to fld.
         ILS/DME 110.7 I-DTW Chan 44 Rwy 03L LOM REVUP NDB
         ILS/DME 110.7 I-DWC Chan 44 Rwy 21R
         ILS 111.5 I-HUU Rwy 03R
         ILS 111.5 I-EJR Rwy 21L
         ILS 108.5 I-DMI Rwy 27 LOM SPENC NDB
         ASR
      COMM/NAVAID REMARKS: Willow Run VOR/DME out of svc indefinitely.
```

Fig. 6-2. *Detroit Metro Airport is an example of dual Ground Control—one frequency for the west half of the field and another for the east.*

Taxiing for Takeoff (Required)

You have ATIS information, have tuned to the correct Ground Control frequency, and are ready to taxi out *now*—not one, three, or five minutes from now. You can roam around the ramp or any uncontrolled area all you want, but don't venture out onto any taxiway until you have contacted Ground and have received permission to taxi.

> **You:** Lincoln Ground, Cherokee One Four Six One Tango at the terminal with Information Lima. VFR Omaha.
>
> *GC:* *Cherokee One Four Six One Tango, taxi to Runway Three Five.*
>
> **You:** Roger. Cherokee One Four Six One Tango.

In this contact, be sure to include your location (e.g., " . . . at the terminal . . ."). If you don't, Ground will always come back with something akin to, "Cherokee Six One Tango, where are you parked?" Avoid this needless request by communicating your position in the first call. By including your destination or direction of travel, you might be directed (winds permitting) to a runway more closely aligned with your departure route.

Soliciting Progressive Taxi Instructions (*Optional*)

You have the ATIS, but you're new to the airport and need taxi instructions. Ground controllers understand that taxiway systems can be confusing to pilots, so they will give *progressive* (step-by-step) taxi instructions, *if* you ask for them:

You: Lincoln Ground, Cherokee One Four Six One Tango at the terminal with Information Lima. VFR Omaha. Request progressive taxi.

GC: *Cherokee Six One Tango, roger. Taxi to the edge of the ramp and hold.*

You: Roger, edge of the ramp and hold. Cherokee Six One Tango.

After proceeding as instructed, come to a stop and *wait* for Ground to come back with your next instructions.

GC: *Cherokee Six One Tango, you are at Taxiway Charlie, turn right onto Charlie. Follow Charlie to Runway Three Zero and hold.*

Again, read back the instructions, do as told, and wait:

GC: *Cherokee Six One Tango, cross Runway Three Zero, turn left at the next intersection onto Taxiway Delta, follow Delta to Runway Three Five and hold. Contact Tower.*

Of course, at a large airport, the series of instructions may be much longer, and if you believe at any point that Ground has forgotten about you, or if you are uncertain of your position, don't be afraid to call Ground. That's why they're there.

Taxiing in after Landing (*Required*)

You've touched down, and the tower has instructed you to "contact Ground point niner." (The tower may give you the complete frequency, as "121.9," but more likely will shorten the instruction to "point niner.") When clear of the active runway—and *past* the hold line—change to the Ground frequency:

You: Lincoln Ground, Cherokee One Four Six One Tango is clear of Runway Three-Five (*or* "the active"). Request taxi to the terminal. [Or you can request progressive taxi instructions as above.]

GC: *Cherokee Six One Tango, taxi to the terminal.*

You: Roger. Cherokee Six One Tango.

Any time that you are moving your aircraft in a ground controlled area, stay tuned to the Ground frequency. Despite the fact that you have been cleared to "taxi to the terminal," Ground may need to give you subsequent instructions.

Example: You've landed at Lincoln, and Ground has authorized you to proceed to the terminal. A moment or so later, another aircraft requests taxi-out permission on the same strip you're occupying inbound. To ensure clearance of the two aircraft, Ground calls you:

GC: *Cherokee Six One Tango, pull to the right at the next intersection to let the Citation pass.*

You: Will do, Cherokee Six One Tango.

Or any other situation could arise warranting instructions to you. If you have turned off your radio, the message will fall on dead ears. Just because you're "cleared to the terminal" doesn't mean the clearance can't be interrupted or changed.

An obvious point? Sure, but pilots have been known to switch everything off—beacons, radios, transponder, etc.—once cleared to the parking areas, and except for the transponder, that's a no-no in the eyes of the powers-that-be.

Moving the Aircraft from One Ground Location to Another (*Required, Except on Ramp/Uncontrolled Areas*)

Let's say you need to taxi to the other side of the field for a minor repair at a radio shop. To do so, you have to use a taxiway and cross the active runway. Permission for both is mandatory:

You: Lincoln Ground, Cherokee One Four Six One Tango at the terminal. Request taxi to King Radio.

GC: *Cherokee Six One Tango, taxi on Foxtrot. Hold short of Runway One Eight.*

You: Roger, hold short of Runway One Eight. Cherokee Six One Tango.

GC: *Cherokee Six One Tango, cleared to cross Runway One Eight.*

You: Roger. Cleared across One Eight. Cherokee Six One Tango.

Getting a Current Altimeter Setting (*Optional*)

You're ready to taxi from the ramp. After listening to the ATIS, however, you find that the altimeter setting given in the recording doesn't coincide with the field elevation. Realizing that the ATIS may be—or is—almost an hour old, you want the current reading. Incorporate your request for this data in the initial call to Ground:

> **You:** Lincoln Ground, Cherokee One Four Six One Tango at the terminal with Information Lima. Ready to taxi. VFR Omaha. Request altimeter setting.
>
> *GC: Cherokee One Four Six One Tango, Lincoln altimeter is three zero zero five. Taxi to Runway Three Five.*
>
> **You:** Roger, three zero zero five. Taxi to Three Five. Cherokee Six One Tango.

For a Radio Check (*Optional*)

You're on the ramp and want to determine the clarity and volume of your transmission. The easy way to do this is to call Ground and ask for a radio check:

> **You:** Lincoln Ground, Cherokee One Four Six One Tango radio check.
>
> *GC: Cherokee Six One Tango, loud and clear.*
>
> **You:** Roger, Cherokee Six One Tango.

If Ground comes back with "transmission is scratchy and volume weak," don't indulge in an explanatory harangue. You've got a sick radio, Ground has told you so, so get off the air, and have a technician look at it. Ground can't help you a bit!

IS THE AIR CLEAR?

Now is as good a time as any to stress the point of listening before you speak. In other words, is the air clear?

You're at the ramp and have just tuned to Ground for taxi permission. The first thing you hear is an IFR aircraft requesting its clearance. Ground responds with:

> *GC: Cessna Three Four Romeo is cleared as filed to Denver. Maintain three thousand, expect ten thousand fifteen minutes after departure. Departure Control one two six point six. Squawk zero three two five. Fly heading two four zero after departure.*
>
> [A brief period of silence]
>
> **34R:** Cessna Three Four Romeo cleared as filed. Three thousand, ten thousand in fifteen. Departure one two six point six, zero three two five, two four zero heading.

> *GC:*   *Cessna Three Four Romeo, readback correct. Taxi to Runway One Eight.*

> **34R:**   Roger, taxi to One Eight. Cessna Three Four Romeo.

As indicated in brackets, there will almost always be a pause, a brief period of silence after Ground has conveyed the clearance, while the pilot is copying the instructions. Then comes the readback.

Don't start transmitting just because the air is momentarily quiet. Recognize what's taking place and give Three Four Romeo time to complete its task and reestablish communications with the controller. This courtesy applies to *all* communications situations, short of an emergency. Give both parties the opportunity to acknowledge instructions, answer a question, repeat an instruction, and the like. Momentary silence doesn't necessarily mean the air is all yours.

Pilots who don't listen and aren't considerate are usually the reason for the squeals and squeaks that distort reception. Two people can't transmit at the same time on the same frequency without creating that discordant cacophony that penetrates the cockpit or headset.

CONCLUSION

Ground Control is the "policeman" for all ground traffic—cars, trucks, tugs, and aircraft. Its use is *mandatory*. Even more than that, however, it is a source of assistance and an overseer of safety, ensuring the smooth flow of ground operations. Additionally, its very existence enables the Tower, responsible for the smooth flow of *flight* operations, to concentrate solely on that responsibility.

It thus behooves all of us to be familiar with what Ground Control can do for us. We must use the service, but we should use it wisely by being clear and concise, and observing the basic rules of courtesy. The Ground Controller will invariably respond in kind.

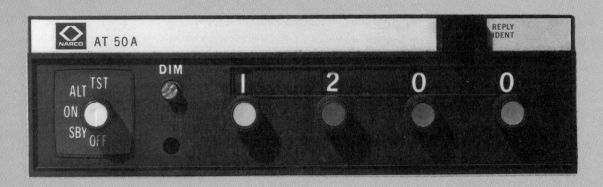

7

Transponders

While we have passingly referred to transponders in some of the previous discussions and examples, this piece of electronic hardware hasn't been a major factor in the various communication processes up to now. Henceforth, however, as we go through the sections on the tower, Approach and Departure Control, and the Air Route Traffic Control Centers, the importance of the transponder increases—as does the importance of the terminology associated with it. Consequently, if you own or rent a transponder-equipped aircraft, some familiarity with it is in order.

THE AIR TRAFFIC CONTROL
RADAR BEACON SYSTEM (ATCRBS)

Simply said, the basic radar system is composed of two elements. One is the *primary* radar, which scans the surrounding area and identifies on the radarscope, or screen, objects such as buildings, radio towers, aircraft without transponders, and aircraft with transponders turned off.

As this relatively minimal identification has safety and traffic control limitations, a *secondary* radar system was developed which incorporates a ground-based transmitter-receiver called an *interrogator*. This system—the Air Traffic Control Radar Beacon System (ATCRBS)—functions in unison with the primary radar and, in the scanning process, "interrogates" each operating transponder. In effect, it "asks" the transponder to reply. The primary and secondary signals are then synchronized and transmit a distinctly-shaped blip or target to the controller's radar screen. That target, however, only tells the controller that there's an aircraft out there with a transponder "squawking" the standard VFR code of 1200. It does not permit more specific identification of the aircraft, which could be important in periods of heavy traffic or poor visibility.

Consequently, each transponder is equipped with an identification feature—the "Ident" button. When the button is pushed, the radar target changes shape to distinguish the identing aircraft from other aircraft on the controller's screen.

Very broadly and non-technically, that is the radar beacon system. For those interested in more detail, there is no substitute for a visit to a radar-equipped tower or an Air Route Traffic Control Center. The specialists in either location are always glad to explain the system and let you watch it in operation—their workload permitting, of course.

THE TRANSPONDER

You'll often hear references to transponder "types" or "modes". In the event there's any uncertainty as to what those references mean, let's take a moment to clear the air.

Several controller-operator decoder modes exist. Modes 1 and 2 are reserved for the military. Mode 3/A is common to both civil and military use. Mode B applies to traffic foreign countries. Mode C identifies a transponder that is equipped with altitude-reporting capabilities. Mode D is not currently in use. Mode S will be the standard in the early 1990s when collision avoidance systems are due to be implemented. (Mode S will automatically transmit the aircraft's N-number, type, and altitude—eliminating the need to dial-in and squawk different codes. It will also permit onboard computer terminals to communicate with ground facilities, resulting in inflight printouts of clearances, weather charts and forecasts, etc.) Currently in civilian flying in the U.S. you'll use Mode 3/A and Mode C.

The transponder illustrated at the beginning of this chapter, (a Narco AT 50A) has five switch positions: OFF, SBY (Standby), ON, ALT (Altitude) and TST. Additionally, there are four code selector knobs so that the pilot can dial in whatever four-digit numerical code Air Traffic Control requests. Each of the four knobs can bring up numbers from 0 to 7, thus allowing for a total of 4,096 separate or *discrete* codes ($8 \times 8 \times 8 \times 8$) hence the frequent reference to a "4096 transponder."

Finally, the transponder is equipped with a small reply light that blinks every time the transponder responds to the radar beacon interrogator. These blinks also confirm to the pilot that the transponder is functioning. Associated with the light is an ident feature. On the unit illustrated the ident button and the reply light are one in the same. On other units the button may be separate. If ATC asks you to "Ident," you merely push the button, and the image on the controller's radar screen changes. If you're one of only a few aircraft in the same general area, the controller can track you fairly easily once you and the others have "idented." It's more difficult for him, though, if the activity is heavy. In these cases, he may ask you to report when over a certain landmark for verification of your position.

What we've said above describes the basic Mode 3/A unit. What Mode 3/A doesn't have is the altitude-reporting capability of Mode C. The 3/A is converted to Mode C merely by adding to the system either an *encoding altimeter* or a *blind encoder*. The first is a normal-appearing altimeter that is coupled to the transponder. As the aneroid bellows expand and contract with pressure changes, those changes are converted to coded response pulses by the transponder, which then transmits the aircraft's altitude (to the nearest 100-foot level). The blind encoder performs the same function, but the unit is usually physically located on the firewall, out of the pilot's sight.

Of the two, the blind encoder has certain advantages over the panel-mounted encoding altimeter. For one, it's usually a little less expensive and can be installed at about the same cost. Also, if it fails, it can be removed for repair and the aircraft operated as usual, except where Mode C is required.

On the other hand, if the encoding altimeter goes out and you still want or need altitude-reporting capability, the entire altimeter has to be removed and the aircraft is grounded pending repairs. Finally, installing an encoding altimeter means the replacement of what is probably a perfectly functioning unit. If you can sell the unit, fine. Otherwise, you've got a good altimeter to put on your fireplace mantle or office desk to impress visitors.

With the concern for inflight safety heightened by recent midair collisions and reported near-misses, more stringent operating and equipment requirements have been established:

- If your aircraft is equipped with a transponder, you must use it (including Mode C, if equipped) whenever you're in controlled airspace. In most areas of the U.S., this means just about everywhere.

- Effective July 1, 1989, a Mode C transponder is required from 10,000 feet MSL to the floor of the Positive Control Area (typically 18,000 feet MSL), except at or below 2,500 feet AGL. (Prior to that date the requirement starts at 12,500 feet MSL.)

- Effective July 1, 1989 a Mode C transponder is required within 30 nautical miles of a Terminal Control Area's (TCA's) primary airport, from the surface up to 10,000 feet MSL. (Depicted on Sectionals by a very thin blue circle.)

- Effective December 30, 1990, a Mode C transponder is required within and above an Airport Radar Service Area (ARSA), up to and including 10,000 feet MSL.

- Effective December 30, 1990, a Mode C transponder is required from the surface to 10,000 feet MSL within a 10 nautical mile radius of certain designated airports (initially Billings, Montana, and Fargo, North Dakota); except below 1,200 feet MSL outside of the Airport Traffic Area.

✈ Aircraft without electrical systems are exempt from many of the requirements, and other exceptions can be made by ATC on a case-by-casebasis.

These requirements will severely limit the airspace available for non-Mode C aircraft near urban areas, but Mode C has a lot of advantages, not the least of which is the added safety it provides. Plus, it has a strong influence on keeping pilots honest. Should you intentionally or accidentally penetrate a TCA without approval, somebody on the ground knows it. Perhaps your particular aircraft can't be identified (if you haven't contacted Approach Control or the tower), but the FAA is developing improved means of tracking airspace violators. If caught, it could mean a license suspension of 60 days or longer. With Mode C in operation, however, you know that you are being, or could be monitored. This should have a certain effect on regulations abidance and attention to what you're doing.

TRANSPONDER CODES

Every pilot should be familiar with the standard numerical codes that are controlled by the four knobs on the set. The most common is the code 1200, which is standard for all VFR altitudes. When using Approach or Departure Control or Center, however, ATC will give you some other code (called a *discrete* code) to "squawk" for specific identification of your aircraft.

Let's amplify that a bit. Say that you're on a cross-country and have contacted an Air Route Traffic Control Center ("Center") for en route VFR traffic advisories. As there could be a lot of 1200s flying around out there, you're told to squawk 2056 (for example). This is the discrete code assigned only to you. The ATC computer recognizes the code and displays your aircraft's ground speed and altitude of Mode C equipped next to your target on the controller's screen. Now he has you distinguished from all other aircraft and is better able to alert you to possible conflicting traffic.

The same principle applies when you're about to enter a TCA, an ARSA or a TRSA (Terminal Radar Service Area). *Before* you enter the controlled area you'll be given a discrete code to squawk.

Barring these situations and the emergency conditions we'll discuss in a moment, put the transponder on 1200 and leave it there.

Emergency Codes

The typical VFR or IFR private pilot need be concerned with only two emergency codes—one for a real emergency, and one in the event of radio failure.

Code 7700 is used exclusively for a bona fide inflight emergency. Engine failure, fire, loss of control, whatever—dial in 7700 as soon as you can. As controllers put it, that code "rings bells and lights lights" on the scope and attracts immediate attention to the aircraft in distress. In effect, it's the same as the verbal

"Mayday" cry for help. ("Mayday", by the way, comes from the French, "M'aidez"—"Help me".) If you're in voice communication with ATC and report an emergency, you may be asked to "Squawk Mayday". This means ATC wants you to dial in 7700.

To indicate radio failure, two codes come into play: 7700 and 7600. When you know your radio has died (transmitter and/or receiver), go to 7700 for one minute, and then change to 7600 for 15 minutes. Repeat the cycle, if necessary. (We'll talk more in Chapter 12 about what happens in the event of radio failure.)

Other Codes

A few other standard but reserved codes are:

0000:   For military use only

4000:   For military use in warning or restricted areas

7500:   For alerting ATC that a hijack is in progress

To summarize the codes, the following might help. Starred* items indicate codes civilian pilots should never use.

Code	Type of Flight	When Used
0000*	Military only	North American Air Defense
1200	VFR	All altitudes unless otherwise instructed
4000*	Military VFR/IFR	Warning/Restricted Areas
7500	VFR/IFR	Hijack
7700	VFR/IFR	Emergency - "Mayday"
7700 (1 min.) 7600 (15 mins.)	VFR/IFR	Loss of radio communications
7777*	Military only	Interceptor operations
Assigned by ATC	VFR/IFR	When using Center or Approach/Departure Control

TRANSPONDER OPERATION

STANDBY: After engine start-up, turn on the radio(s) and put the transponder switch to the SBY position. This allows the set to warm up without replying to the radar interrogator. Keep the switch in this position until cleared for takeoff or when you have actually begun the takeoff roll.

ON or ALT: With Mode 3/A, switch from SBY to the ON (or in some units, NORMAL) position when cleared for takeoff, and leave it there throughout the flight, unless directed otherwise by ATC. If the unit has Mode C, switch to the ALT position. This not only turns the set on, but now you'll also be reporting your altitude. As mentioned earlier, if you have a transponder when operating in controlled airspace (unless ATC directs otherwise), the transponder *must* be switched to ON (Mode 3/A), or, if equipped with an altitude encoder, ALT (Mode C).

OFF: Turn the transponder off as soon as you have landed, either on the rollout or when you're clear of the runway. If you leave it on, all you're doing is painting an enhanced image on the screen because of your proximity to the radar beacon antenna.

IDENT: Only when ATC asks you to "Ident" do you push the IDENT button—and just once. The signal sent will change the shape of the blip on the screen. Now the controller can more readily identify your aircraft and its relation to ground obstacles and other airborne traffic.

TERMINOLOGY

Finally, the terminology, or phraseology, associated with Mode 3/A and Mode C transponder operation:

Squawk: The origin of this somewhat odd term goes back to World War II and a radar beacon system called IFF (Identification, Friend or Foe). Allied aircraft were equipped with "transmitters" which replied to radar sweeps with a sound similar to a parrot's squawk. Today the term is used by controllers and pilots alike to indicate that the transponder is on and that a certain four-digit code should be, or has been, dialed in. If a controller asks you to "squawk two zero five six" (or any code), he wants you to enter those digits in the transponder. That is *not* an instruction to "ident" however. If you're squawking a code other than 1200 and are told to "squawk VFR" it means to change the code back to 1200.

Ident: When a controller says, "Cherokee Six One Tango, ident," he wants you to push the little IDENT button so that he can more positively identify your aircraft. But just push the button only once and only momentarily. Usually, but not always, after the radar target changes shape, the controller will come back to you with, "Cherokee Six One Tango, radar contact." Then it's appropriate to acknowledge with, "Roger, Cherokee Six One Tango." On the other hand, when asked to ident, you don't have to reply with, "Roger, Cherokee Six One

Tango identing.'' Just push the ident feature and say nothing. The blip change on the screen is acknowledgment enough for most controllers.

Occasionally, ATC may come back with, "Cherokee Six One Tango, I did not receive your ident. Ident again." In this case, acknowledgment is in order: "Roger, Cherokee Six One Tango." Now repeat the ident.

Squawk (number) and Ident: This asks you to insert a certain code and push the IDENT button or light. The word "and" may or may not be included in the instruction.

Stop Squawk: When you hear this, it means that the controller wants you to turn the transponder to OFF.

Squawk Standby: This is an instruction to turn the switch from ON or ALT to SBY—the standby position. Remember, the transponder is not off. It's still warm but isn't responding to any interrogation and thus not transmitting. Your aircraft will be reflected on the screen only by the primary radar—not the secondary. (You might recall the pilot in Cherokee 41966 back in Chapter 1. He switched the transponder to STANDBY, not understanding that the controller wanted him to squawk 0252 and then *stand by* for further instructions.)

Stop Altitude Squawk: If you have Mode C and hear this, merely switch from the ALT to the ON position. You're now functioning in Mode 3/A only, which provides aircraft identification but not altitude.

Squawk Mayday: This we mentioned earlier. You've verbally communicated an emergency to ATC, and for positive identification, the controller wants you to change your code to 7700.

SOME REMINDERS AND TIPS

Remember to put the transponder in the SBY position after engine start. Change it to ON or ALT only after receiving takeoff clearance or during the takeoff roll. Leave it on throughout the flight, unless directed otherwise. Turn it off as soon as you have landed and can do so safely.

In changing codes, *always* avoid even a momentary display of 7500, 7600, and 7700. These codes are strictly for radio loss or an emergency.

When changing from one code to another, repeat the new code back to the controller, and then make the change immediately. It's also a good idea to write down the new code the controller has given you so that you won't have to sheepishly ask, "What was that squawk you wanted?"

If you have a transponder (with or without Mode C), make it known to ATC in your initial contact. The call should go like this:

Turner Tower, Cherokee One Four Six One Tango over Hollow Lake at three thousand five hundred, squawking one two zero zero, with Information India.

Nowadays, controllers shouldn't have to ask, "Are you transponder equipped?" If you do have a transponder, it *must* be turned on whenever in controlled airspace. And if you are squawking the wrong code, the controller will know it.

CONCLUSION

The transponder is an important element in air traffic control and the ever increasing drive for safety and becomes even more essential as new FAA regulations go into effect.

The nice thing about a transponder, though, is that it doesn't require a lot of pilot expertise. In some respects, it's a little like ATIS: it does more for you than you have to do yourself. With ATIS, you just sit and listen. With the transponder, you turn it on, understand the limited phraseology, do as you're asked, and that's it. It—not you—keeps the people on the ground informed of where you are, and, if Mode C-equipped, your altitude. Functioning as it does, it reduces the need for voice communications and contributes to the safety of all of us.

But, a final word is essential. The transponder and what it does should never be allowed to lull you into complacency. Despite the controller's skill and the sophistication of his electronic equipment, it's still the pilot's job to see and avoid. IFR aircraft in instrument meteorological conditions within controlled airspace are assured of horizontal and vertical separation from all other aircraft. As a VFR pilot, however, the most you'll usually get are advisories of the positions of other aircraft and safety alerts when known conflicts seem imminent. In a TCA the control is greater. Otherwise, it's up to you, with whatever help ATC can give you, to do what you always should be doing—protecting your own skin. The final responsibility for that sits in the left seat of every airplane.

8

The Control Tower

Ground Control has directed you to the active runway (not literally *to* it, but to the runup or hold area adjacent to it). You've completed the pre-takeoff check and are ready to switch frequencies to the glassy greenhouse that surveys the ground and air activity surrounding it. Before proceeding further, a simple question:

WHAT DOES THE TOWER DO?

Perhaps the question is so basic that the answers are obvious. Regardless, let's pursue the obvious anyway.

The tower controls all local aircraft traffic within what is called the *Airport Traffic Area* (ATA):

✈ To land at or depart from a tower-controlled airport, you *must* maintain two-way communications.

✈ To cross through a portion of the ATA, you must advise the tower of your intentions and follow the vectoring or directions of the controller.

✈ You must do what the tower tells you, unless those instructions would place you or your aircraft in jeopardy. Even then, however, you are required to advise the tower of alternate actions you are taking or plan to take. The tower is not an advisory service, such as UNICOM or Flight Service. It's a *controlling* agency whose directions must be followed, except in an emergency.

(The term "Airport Traffic Area" may change to "Control Tower Area" if an FAA proposal is adopted.)

AIRPORT TRAFFIC AREAS AND CONTROL ZONES: DIFFERENCES, SIMILARITIES, REQUIREMENTS

In terms of radio requirements, there is apparently a need to clarify the differences and similarities of Airport Traffic Areas (ATAs) versus Control Zones (CZs)—if the incidents of unannounced aircraft popping up in the traffic pattern and unnecessary communications from a pilot at 7500 feet in CAVU weather requesting permission to transit a Control Zone are any criteria. So at the risk of boring the knowledgeable, bear with us as we go back to basics.

Airport Traffic Area

The existence of an airport traffic area is indicated on the Sectional, only by the blue tower-controlled-airport symbol:

- ATAs exist *only* at those airports where there is a control tower in operation. When the tower is closed down, for example, from 2200 hours to 0600 hours at some airports, the ATA does not exist.

- The ATA is always a five-statute-mile radius from the airport and extends vertically up to, but not including, 3000 feet AGL. Remember, that's *AGL*—not MSL.

- There are no weather minimums unique to operating within an ATA.

- Radio communications must be established to enter, depart, or operate within the five-mile radius below 3000 feet AGL.

- Radio contact is *not* required if overflying the ATA (i.e., at or above 3000 feet AGL).

Control Zone

Do not confuse Control Zones with Airport Traffic Areas. Whereas ATAs are not graphically depicted on Sectionals, CZs are—they are the broken blue lines, either circular or keyhole-shaped, surrounding many airports. CZs exist solely to provide extra protection for IFR operations at those airports. Most CZ airports have a tower (and thus an ATA also) or an FSS. There are no communications requirements associated specifically with a CZ *except* if you want to operate within a CZ under Special VFR (less than basic VFR weather minimums—worse than 3 miles visibility and 1000-foot ceiling). At a few of the busiest airports Special VFR is never permitted; this is indicated by blue **T**'s replacing the normal broken blue CZ line.

Here are a few points to remember about a Control Zone:

- ✈ It usually coincides with the five-mile radius of the ATA, except for the IFR extensions.

- ✈ Although the lateral limits of a circular CZ often correspond to the lateral limits of an ATA, as opposed to the ATA, the CZ rises vertically to 14,500 feet MSL, the base of the Continental Control Area.

- ✈ When the field is at or above basic VFR weather minimums, radio communication is *not* required to fly through any part of a CZ that is not within the ATA (that is, any part of the CZ beyond the 5-SM radius and/or 3000 feet AGL and above). To clarify that point, refer to FIG. 8-1. When you are within the shaded area, you are within the CZ but outside or above the ATA. While you would be wise to notify the tower of your existence and position, there is no requirement to do so under basic VFR.

- ✈ When the field is below basic VFR, clearance from either the tower or Flight Service is required prior to operating within the CZ.

Control Zone and Airport Traffic Area Combined

When a CZ and ATA exist at the same airport, they will overlap and yet retain the separate requirements and characteristics outlined above. FIGURE 8-2 is an example of such an airport.

Airports With a Control Zone But No ATA

If you check almost any Sectional, you'll find a plethora of airports colored in magenta, signifying there is no control tower—hence no ATA. Around some of these fields, however, are the broken lines identifying Control Zones. This is not a botch job by the printer. A CZ exists when one or a combination of the following is located on the field and can provide the essential weather services:

- ✈ an operating tower and a qualified weather observer

- ✈ an operating Flight Service Station

- ✈ a National Weather Service Office

- ✈ a qualified weather observer

One example is Russell, Kansas (FIG. 8-3). Even though without a control tower, it has a CZ because a Flight Service Station, and thus a weather observer,

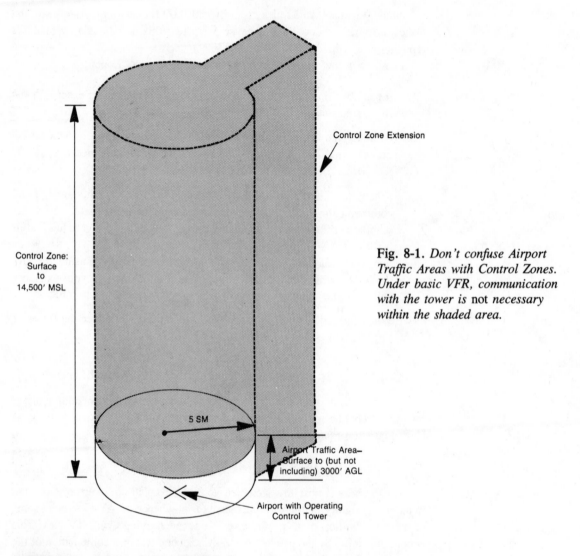

Control Zone Extension

Control Zone:
Surface
to
14,500' MSL

5 SM

Airport Traffic Area—
Surface to (but not
including) 3000' AGL

Airport with Operating
Control Tower

Fig. 8-1. *Don't confuse Airport Traffic Areas with Control Zones. Under basic VFR, communication with the tower is* not *necessary within the shaded area.*

Fig. 8-2. *Ardmore, Texas, has both a Control Zone, depicted by the broken blue keyhole-shaped line, and an Airport Traffic Area, not depicted graphically, but indicated by the presence of an operating control tower. In this case, NFCT means non-federal control tower.*

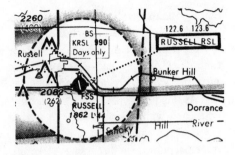

Fig. 8-3. *Russell has no tower, but it has a Control Zone because a Flight Service Station is located on the airport.*

is on the premises. Another example (FIG. 8-4) is Shenandoah Valley Airport in Virginia—no tower, and the FSS is some 90 miles away in Leesburg. So what justifies the CZ? Simply the fact that a qualified weather observer is on the field. How do you know this from the Sectional? You don't—directly. By deduction, however, you can conclude it to be a fact: there's no tower, no FSS, and the chances that the National Weather Service would hae an office there are remote. Ergo, there must be a qualified weather observer.

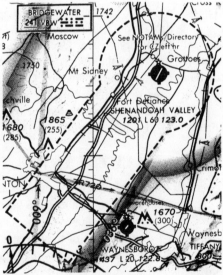

Fig. 8-4. *Shenandoah Valley has no tower and no FSS, yet the Sectional shows a Control Zone. A qualified weather observer is on duty during the CZ's effective hours. Check the A/FD and NOTAMs for those hours.*

If the conditions are VFR at Shenandoah Valley, there is no *requirement* to contact anybody before landing or taking off, despite the obvious safety precaution of keeping others informed of your presence and intentions via the CTAF (UNICOM or 123.6).

On the other hand, as soon as conditions drop below basic VFR minimums, a Special VFR clearance must be obtained through the responsible Flight Service Station before you can enter the Control Zone or depart from it. (Keep in mind that we're talking about a non-tower airport. If a tower were on the airport, that would be the source to contact for the clearance.)

At airports where the tower, FSS, or weather observer is on duty only part-time, the CZ is also part-time and is effective only during the hours listed in the *Airport/Facility Directory*, and NOTAMs.

At this point, if you are still a bit confused and wondering "what do I do when and with whom," refer to FIG. 8-5.

Specific examples of radio contacts are suggested later in this chapter. The principal purpose of discussing ATAs and CZs here is to try to clarify the meaning of each and the conditions under which communications with the tower or Flight Service are essential.

DETERMINING THE TOWER FREQUENCY

For the VFR pilot, two sources are available to determine tower frequencies—the Sectional chart and the *Airport/Facility Directory (A/FD)*.

The Sectional always publishes the frequencies in two locations: adjacent to the airport name, with "CT" preceding the frequency (FIG. 8-6); and on the reverse side (usually) of the "Legend" flap (FIG. 8-7).

The *A/FD* is published six times a year, each issue having an approximately two-month validity period. One advantage of this source is that, for a given location, it gives all the frequencies: UNICOM, Flight Service (either on the field or remoted), Approach/Departure Control, ATIS, Ground Control, Clearance Delivery, and tower (FIG. 8-8). The Sectional is never this complete.

Just be sure you always use the *current* editions of the Sectional and *A/FD*. Frequencies do change.

DOING WHAT THE TOWER TELLS YOU

Yes, you are the pilot in command, with certain responsibilities and authority. And, yes, the tower is there to serve you, along with all the other pilots in the area. Neither condition, however, alters the fact that *no* pilot has the right to go his or her way regardless of the tower's instructions. To repeat what we said a moment ago, unless an emergency suddenly develops or adherence to an instruction would violate an FAR, you must comply with the tower's directives.

The controller is spacing, separating, coordinating, and overseeing all the traffic within his area of responsibility. He has a plan to get everyone up or down with minimum delays. He can't have the whole operational pattern disrupted because some character does a 360 on the downwind leg or chooses to land on Runway 21 when Runway 18 is the active runway. Perhaps judgment decrees that a 360 is essential for spacing or safety. If so, advise the tower *before* starting the maneuver. Perhaps 21 is a better runway because of the wind. Fine, but get permission before you switch to another pattern.

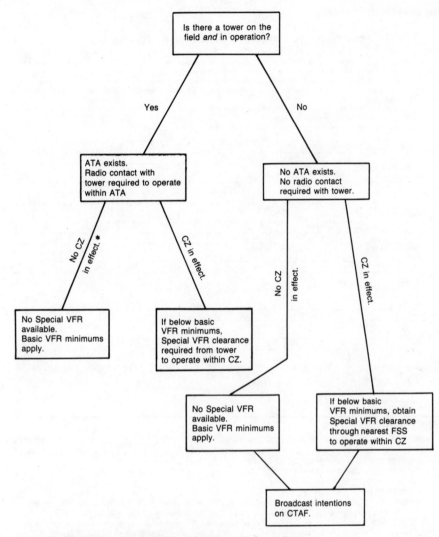

Notes: Special VFR is not available within some high-activity Control Zones.
Control Zone effective hours are listed in the *A/FD*.
★ Tower airports without CZs are extremely rare, but do exist.

Fig. 8-5. *This diagram will help in determining the communications requirements at airports with ATAs and/or CZs.*

Fig. 8-6. *The most convenient place to find the tower frequency is next to the "CT" on the Sectional.*

CONTROL TOWER FREQUENCIES ON CINCINNATI SECTIONAL CHART

Airports which have control towers are indicated on this chart by the letters CT followed by the primary VHF local control frequency. Selected transmitting frequencies for each control tower are tabulated in the adjoining spaces, the low or medium transmitting frequency is listed first followed by a VHF local control frequency, and the primary VHF and UHF military frequencies, when these frequencies are available. An asterisk (*) follows the part-time tower frequency remoted to the colocated full-time FSS for use as Airport Advisory Service (AAS) during hours tower is closed. Hours shown are local time. Radio call provided if different from tower name.

Automatic Terminal Information Service (ATIS) frequencies, shown on the face of the chart are normal arriving frequencies, all ATIS frequencies available are tabulated below.

ASR and/or PAR indicates Radar Instrument Approach available.

CONTROL TOWER	RADIO CALL	OPERATES	ATIS	FREQ	ASR/PAR
BENEDUM	CLARKSBURG	0700-2300		126.7 252.9	ASR
BLACKSTONE AAF-PERKINSON	BLACKSTONE	0700-1700 MON-FRI APR-SEP		126.2	
BLUE GRASS	LEXINGTON	0600-2400	126.3	119.1 257.8	ASR
BOLTON NF		0700-2100		119.0	
CHARLOTTESVILLE-ALBEMARLE	CHARLOTTES-VILLE	0648-2300		124.5 242.7	
CINCINNATI-LUNKEN	LUNKEN	0700-2300	120.25	118.7* 257.8	
COX-DAYTON INTL	DAYTON	CONTINUOUS	125.8	119.9 257.8	ASR
EASTERN WV REGIONAL/SHEPHERD	MARTINSBURG	0830-1630 TUE-FRI O/T SUBJECT TO ANG		124.3* 236.6	
GREATER CINCINNATI INTL	CINCINNATI	CONTINUOUS	135.3	118.3 286.6	ASR
GREENBRIER VALLEY NF	LEWISBURG	0900-1900		118.9	
GREENSBORO-HIGH POINT-WINSTON SALEM REGIONAL	GREENSBORO	CONTINUOUS	128.55	119.1 226.3	ASR
LYNCHBURG-GLENN	LYNCHBURG	0630-2300	119.8	120.7 257.8	
MORGANTOWN-HART	MORGAN-TOWN	0700-2300		120.0* 257.8	
OHIO STATE UNIV	OHIO STATE	0700-2300	121.35	118.8 258.3	

Fig. 8-7. *Another place to find tower frequencies is on the reverse side (usually) of the Legend flap of the Sectional.*

CLARKSBURG

§ **BENEDUM** (CKB) 0 NE UTC–5(–4DT) 39°17′44″N 80°13′44″W   **CINCINNATI**
    1203   B   S4   **FUEL** 100LL, JET A   OX 1, 2   CFR Index A   H-4I, 6H, L-22F, 23D, 24E
    RWY 03-21: H5198X150 (ASPH-GRVD)   S-70, D-90   HIRL   0.4% up NE   **IAP**
      RWY 03: REIL. VASI(V4L)—GA 3.44°TCH 59′. Trees.   **RWY 21:** MALSR. Ridge.
    RWY 13-31: H2500X75 (ASPH)   S-26
      RWY 13: Ridge.   **RWY 31:** Trees.
    AIRPORT REMARKS: Attended continuously. CLOSED to unscheduled air carrier ops with more than 30 passenger seats except PPR call arpt manager 304-842-3400. Rwy 13/31 CLOSED to air carrier ops. Rwy 21, for MALSR, 0300-1100Z‡, key 126.7-3 times for low, 5 times for medium, 7 times for high. Lights remain on for 15 minutes Deer on and in vicinity of arpt. Ldg fee for all acft over 6500 lbs. Control Zone effective 1200-0400Z‡.
    WEATHER DATA SOURCES: LAWRS (304)842–4465.
    COMMUNICATIONS: CTAF 126.7   **UNICOM** 123.0
      MORGANTOWN FSS (MGW) LC 622-2611 NOTAM FILE CKB
      CLARKSBURG RCO 122.1R 112.6T (CHARLESTON FSS)
    Ⓡ **CLARKSBURG APP/DEP CON** 119.6 (West) 121.15 (East) (1200-0400Z‡)
    Ⓡ **CLEVELAND CENTER APP/DEP CON** 125 1 (0400-1200Z‡)
    CLARKSBURG TOWER 126.7 (1200-0400Z‡)   **GND CON** 121.9
    STAGE II SVC ctc APP CON
    RADIO AIDS TO NAVIGATION: NOTAM FILE CKB.
      CLARKSBURG (L) VOR/DME 112.6   CKB   Chan 73   39°15′11″N 80°16′05″W   042°2.6 NM to fld.
      1430/04W.
      ILS 109.3 I-CKB Rwy 21 Glide Slope unusable below 1600′. ILS unmonitored when twr clsd.
      ASR

Fig. 8-8. *The* Airport/Facility Directory, *still another source of tower frequency information—and a whole lot more. Here, the A/FD reveals that the tower is open only between the hours of 1200-0400Z.*

Does any of this sound fundamental? Undoubtedly, but such unannounced or unapproved deviations from instructions are not that rare. Stick around a controlled airport for a while. Listen and observe. You'll see.

WHAT IF YOU DON'T UNDERSTAND?

Some controllers speak rapidly; some occasionally slur their words; some may use a term or issue an instruction you don't understand; perhaps your radio reception is fuzzy; maybe someone cuts in just as the tower is telling you what to do and all you hear are squeals and squawks. Whatever the case, you don't understand the controller's message.

Here is where uncertainty can have unpleasant consequences. If you don't understand, don't just "roger" the instructions and hope that whatever you do will be the right thing.

One of the basic elements of communications is *understanding*. The word "communication" comes from the Latin *communicare,* meaning "to share; to make common." When the words that flow between speaker and receiver are understood—when the receiver has the same idea in his mind that the speaker had in his—there is a sharing, a commonality. Hence, there is *communication.*

The reasons for breakdowns in human relations, in business transactions, and in international relations are myriad. Perhaps the one factor most responsible is the failure of people—or nations—to communicate in ways that nurture mutual understanding. It's too easy to *assume* that what I say has the same meaning to you that it has for me. Messages are distorted by wishful hearing, wandering attention, mistrust, and word choice (i.e., jargon, slang, unfamiliar terms, and imprecise words).

To an Englishman, a "bonnet" means the hood of his car, while Americans typically think of baby hats or the Fifth Avenue Easter Parade. A "trolley" is a restaurant serving cart in Britain, but Americans conjure up thoughts of San Francisco's Powell Street and electrified streetcars that run on tracks. Back home, the tower tells you to "fly the final"—perfectly clear words, and you say to yourself, "What the hell!" I'm flying, and I'm *on* the final. What's he talking about?" He means to *extend* the final by making S-turns for better spacing. Communication is lost in the haze of jargon.

FIGURE 8-9 illustrates part of the ongoing communication problem: It's a long way from A to D, and the road is frought with many unintended detours.

Compounding the inherent causes of poor communications is the very human tendency to try not to appear stupid. Most people are reluctant to admit they didn't understand a directive or an instruction, especially if they feel the communicator expects instant comprehension. It's a matter of preserving one's self-esteem, of not losing face. One of the more meaningless questions an instructor, a boss, or a parent can ask is "Do you understand?" Unless complete trust between the

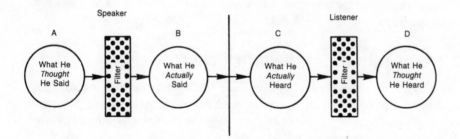

Fig. 8-9. *The communication problem.*

two parties exists, not many people are willing to say, "I didn't understand a thing you said." That's admitting ignorance or—worse—stupidity.

Pilot and controller talk to each other on a one to one basis. There is, however, an unseen listening audience out there capturing everything the two are saying. Well aware of this, the pilot who is uncertain about radio communications faces the added concern of exposing his uncertainty to his airborne compatriots. So he "rogers" the instructions and prays that everything will work out all right.

Our point here is to air the concerns every pilot has—or had—about radio procedures and communications. If you've got a few thousand hours in your log, the concerns have (hopefully) evaporated. The less experienced you are, the greater the concerns are likely to be—which, of course, is one reason why those of us who learned to fly at uncontrolled airports venture most hesitantly into the larger aerodromes.

As we stressed in an earlier chapter, know what you're supposed to say, and then practice it—even at home with a tape recorder. That will sharpen your "sending" ability. Then, while in the air, listen long and hard to what the tower is telling other pilots. If you've never used Approach or Center, tune in to the appropriate frequency and do some eavesdropping. Just getting the *feel* of the verbal exchanges will help.

If, despite all your practice, you still don't understand what the tower tells you, ask for immediate clarification:

"Tower, say again."
"Tower, Cherokee Six One Tango. Say again more slowly."
"Your transmission was garbled. Say again, please."
"Am unfamiliar with the term. Please explain."
"You were cut out. Please repeat instructions."
"Am unfamiliar with the area. Please identify location of reporting point."
(The tower had told you to report "over the twin stacks," but where are the "twin stacks"?)

The controller is similar to a sports coach, and the pilots on the ground or in the air are the players. The controller is calling the plays. He's in charge. The players are expected to do what he says. If a "play" won't work, the "coach" should be notified. If there is misunderstanding or confusion, it had better be cleared up *now*, because the "game" in the air is far more consequential than any earthbound contest.

Yes, there is a "team" relationship between pilot and controller. Without clear communications between them, potential problems are invited. Working together in harmony, each will be a winner.

THE TAKEOFF CONTACT

The start of this chapter had you ready for takeoff. Let's continue from there.

Assume that you're Number One to go and are still in the runup area, some 75 feet or so from the hold line. After the pre-takeoff checks have been completed, taxi to the hold line and come to a complete stop. At this point, you're clear to switch from Ground Control to the tower frequency. Before making the call, however, check your volume, and if you have two navcoms, be sure you're ready to transmit as well as receive on the proper frequency.

Why call at the hold line (FIG. 8-10) and not when you have completed your runup? Suppose that an aircraft is on a relatively short base about to turn final. If you're still back in the runup area, you probably won't have enough time to get to the runway before the landing craft is on your tail. Hence, the tower might tell you to "taxi closer and hold." But if you're at the hold line, the tower might be able to clear you for an immediate takeoff. Through this simple maneuver, traffic is expedited and aircraft on the ground behind you have less idling and fuel-burning time.

When you're *really* ready to go, the call is nothing more than:

You:	Cedar Tower, Cherokee One Four Six One Tango ready for takeoff. Request east departure.
Twr:	*Cherokee Six One Tango cleared for takeoff. East departure approved.*
You:	Roger, cleared for takeoff, Cherokee Six One Tango.

Remember KISS. *No need* for something like this:

You:	Cedar Control Tower, this is Cherokee One Four Six One Tango, Over.
Twr:	*Cherokee One Four Six One Tango, Cedar Tower.*
You:	Cedar Control Tower, this is Cherokee One Four Six One Tango. We're ready to take off on Runway Three Six. We would like to leave the pattern and depart to the east. Over.

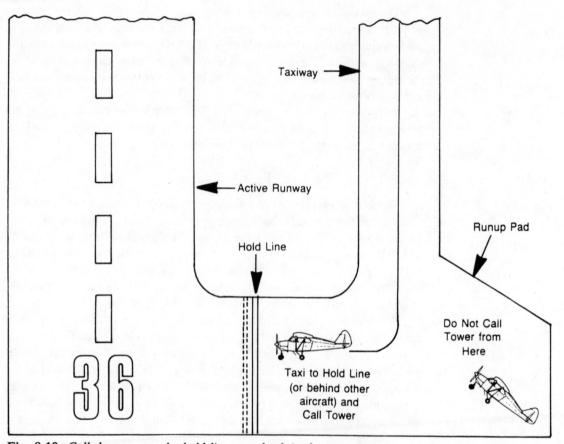

Fig. 8-10. *Call the tower at the hold line—not back in the runup area.*

Twr: *Cherokee Six One Tango cleared for takeoff. East departure approved.*

You: Roger. Six One Tango. East departure approved.

That transmission consumed 52 words, not counting the tower's replies. What have you said that you didn't say in 22 words in the other departure call? Nothing. And the tower knows you're going to use Runway 36, so why state the obvious?

Now let's assume that you've made the call and the tower acknowledges with: "Cherokee Six One Tango, hold for landing traffic." Your reply should indicate your understanding and compliance: "Cherokee Six One Tango holding."

The landing aircraft has landed, and the tower contacts you again: "Cherokee Six One Tango, taxi into position and hold." Your reply: "Position and hold, Cherokee Six One Tango."

You haven't merely "rogered" the instruction. You've *confirmed* to the tower that you have understood and will comply with the instructions.

When the landing craft has cleared the runway, the tower calls you once more: "Cherokee Six One Tango, cleared for takeoff." What's missing? Should you take off under the possibly false assumption that your eastbound departure has been approved? *No.* You might start your run, but don't assume anything. Contact the tower again:

You:	Tower, Cherokee Six One Tango. Is eastbound departure approved?
Twr:	*Cherokee Six One Tango, roger, eastbound departure approved. Remain north of the buildings.*
You:	Wilco ["will comply"], Cherokee Six One Tango.

Another situation: after your initial "ready for takeoff" call, the tower says: "Cherokee Six One Tango cleared for immediate takeoff."

What's the controller saying? He's telling you to get out on that runway and *go.* "Immediate" means *immediate.* You don't idle your way to 36 and take your sweet time setting everything up. Taxi out as rapidly and safely as you can to the center of the runway, line up the airplane, and apply power *now.*

This sort of instruction doesn't require your acknowledgment. The fact that your aircraft is moving is indication enough that you have received the message. If you sit on the runway in takeoff position with no visible activity on your part, you're certain to hear from the man or woman in the elevated enclosure.

A variation of this situation: Responding to your ready-to-go call, the tower comes back with: "Cherokee Six One Tango cleared for immediate takeoff or hold short." It's decision time. If you're *really* ready, taxi out and get going; the rolling aircraft confirms your intentions. Otherwise, tell the tower what you're going to do. Silence leaves the controller in a state of uncertainty—which makes controllers very unhappy. Don't play *I've Got a Secret.* Quickly reply, "Cherokee Six One Tango will hold."

One other example of these takeoff contacts: Ahead of you at the hold line are two other aircraft. You've completed your run-up and are ready to go, but the other planes are just sitting there. If, for example, they're waiting for IFR clearances, they may remain sitting for some time, while you stay patiently in line burning up fuel. So what do you do? The logical thing is to call the tower. If possible, the tower will clear you ahead of those who precede you:

You:	Cedar Tower, Cherokee One Four Six One Tango, number three in sequence, ready for takeoff. Request east departure.
Twr:	*Cherokee Six One Tango. Taxi around the Cessna and Mooney. Cleared for takeoff. East departure approved.*
You:	Roger, cleared for takeoff, Cherokee Six One Tango.

CHANGING FREQUENCIES AFTER TAKEOFF

For a variety of reasons, you may want (or need) to change frequencies shortly after taking off. Perhaps you're going to a nearby airport and want to monitor the traffic; your departure route takes you over another airport and clearance to cross it is required; you want to contact Departure Control; you need to open or modify your VFR flight plan with Flight Service; and so on. Whatever the reason, always ask for and receive approval from the tower to change to another frequency *as long as* you are still within the Airport Traffic Area—the five mile radius. The request is: "Cedar Tower, Cherokee Six One Tango requests frequency change."

In all likelihood, the tower will approve the request. If local traffic is heavy or visibility limited, the controller may want you to stay with him until you are well clear of his area. Depending on the circumstances, his response might be: "Cherokee Six One Tango, frequency change approved," or "Cherokee Six One Tango, remain on this frequency. I'll have traffic for you," or "Cherokee Six One Tango, stay with me until clear of the area."

The point is: don't leave the tower while within the ATA until the tower has approved the change. The controller might need to contact you for any number of reasons, and you are still within the airspace that is under his control.

TAKEOFF WITH CLOSED TRAFFIC

This time, you want to sharpen your landings with a few touch-and-gos:

You: Cedar Tower, Cherokee One Four Six One Tango, ready for take-off. Request closed traffic.

Twr: *Cherokee One Four Six One Tango roger. Cleared for takeoff. Closed traffic approved.*

From this point on, you usually won't have to initiate any further contacts with the tower. This is a controlled airport, and the tower knows your intentions. All that's necessary is to acknowledge the controller's transmissions to you:

Twr: *Cherokee Six One Tango, cleared for touch-and-go, Runway Three Six.*

You: Roger. Cherokee Six One Tango.

Again, however, keep the tower informed. When you decide that you've had enough practice, it's a good idea on the downwind leg to tell the tower that this will be a full-stop—even before the controller clears you for another touch-and-go. Initiating the call early on the downwind may help the controller space other aircraft that are either landing or taking off:

You: Cedar Tower, Cherokee Six One Tango, will be full stop this time.

Twr: *Cherokee Six One Tango, roger. Cleared to land, Runway Three Six.*

Whether you initiate the call or it's in response to the controller's clearance for another touch-and-go, be sure to inform him of your intentions. Controllers don't like surprises. However, unlike at MULTICOM or UNICOM airports, do *not* call the tower on downwind or when turning base and final, unless there is a special reason to do so. The controller knows what you're doing, and these calls only clog the air.

APPROACHING-THE-AIRPORT CONTACTS

You're on a cross-country flight and are nearing your destination airport. Just to make it simple, let's say that you're below the TCA (Terminal Control Area), if there is one, and the field does not have a TRSA (Terminal Radar Service Area) or ARSA (Airport Radar Service Area). Be a good Scout before making the initial contact: be prepared!

An airline captain told this story of what came over the air one time. As he was approaching a field, he heard a charter carrier talking to the tower. With names and locations changed to protect the guilty, the call went something like this:

Pilot: Mayflower Tower, this is Rocky Charter Three Niner Niner Uniform.

Tower: *Rocky Charter Three Niner Niner Uniform, go ahead.*

Pilot: Tower, Three Niner Niner Uniform, we're over . . . [to the first officer, with an open mike] Where are we, Harry? [Pause] We're over North Centerville for landing at Mayflower.

Tower: *Roger, Three Niner Niner Uniform. Say your altitude.*

Pilot: Altitude is . . . What's our altitude, Harry? [Pause] Altitude is seven thousand five hundred.

Tower: *Roger, Three Niner Niner Uniform. What's your airspeed?*

Pilot: Let's see, airspeed . . . how fast we going, Harry? [Pause] 375 knots, Tower.

Tower: *What are you squawking, Three Niner Niner Uniform?*

Pilot: We're squawking . . . Harry, what are we squawking?

At this point, the tower broke in:

Tower: *Three Niner Niner Uniform, would it be all right if we talked to Harry?*

Not having personally heard this exchange, we can't vouch for its authenticity. The airline captain, however, left no doubt in our minds that it was almost a verbatim copy of what transpired.

Know what you're going to say, and if you're a little uncertain, rehearse it to yourself before you pick up the mike and press the button. Also, if you've set things up properly, you've listened to the ATIS and monitored instructions to other aircraft on the tower frequency. With these transmissions, you know the winds, altimeter setting, active runway, and traffic pattern direction. Now you're equipped to plan your approach because you have a good idea what the tower will tell you when you establish contact.

At what point should you make the initial tower call? To repeat, it *must* be made *before* you enter the five-mile ATA radius. To the tower personnel, you're a stranger, and they don't appreciate strangers who bounce into their territory unannounced. A reasonable rule of thumb is to introduce yourself about 15 miles out or at the little red flags on the Sectional that identify the various visual reporting points.

After listening to the last transmission to be sure the air is dead, follow this sequence in the call:

1. Tower name.
2. Aircraft type.
3. N number.
4. Position.
5. Altitude.
6. Squawk (if transponder-equipped)
7. ATIS identification.

The initial call would go like this:

Cedar Tower, Cherokee One Four Six One Tango over Grand Lake Dam, level at three thousand five hundred, squawking one two zero zero with Information Lima.

The tower's response may vary, but if everything is normal, you'll soon hear the controller say:

Twr: *Cherokee One Four Six One Tango, enter left base for Runway One Eight.* [plus winds and altimeter if there's no ATIS]

You: Roger, left base for One Eight. Cherokee Six One Tango.

Once again we stress the importance of tersely repeating the tower's instructions. If you're told to "enter left base for One Eight," confirm your understanding with "left base for One Eight."

If the instruction is to "cross midfield at two thousand three hundred for left downwind," something like this will ensure that you're both on the same page in the same hymn book: "Roger, midfield at two thousand three hundred, left downwind."

You're instructed to "report two miles east of the (airport, buildings, city, bridge, water tower, power plant, or whatever)." Your reply is: "Roger, report two east (of whatever). Cherokee Six One Tango."

These responses are more communicative than a good old "roger." Theoretically, "roger" means that you have understood and will do what you've been told. But *do* you understand? Probably, but the tower can't be sure. Of course we don't want to clutter up the airwaves with needless commentaries, but suppose that part of the tower's transmission was garbled or indistinct and you think the runway is 24 when it is actually 34. To plan your pattern accordingly could have disruptive effects. This sort of confusion is unlikely if you monitored ATIS and the tower beforehand, but confusion is still possible.

As just one example, we heard a Cessna 172 not long ago that was piloted by Lieutenant Confusion or Captain Stupid. It was coming into the Kansas City Downtown Airport from the south, with 01 as the active runway. After we heard the Cessna first on Approach Control, we switched to the tower to announce our position (we were not using Approach, just eavesdropping). A few minutes later, the Cessna turned up well north of the airport and hesitantly contacted the tower. The tower clearly identified the active as Runway 01, with right-hand traffic. The Cessna dutifully rogered the information. Next, the tower told the Cessna to fly a heading of 190 degrees on the downwind. Again, the Cessna rogered. Apparently, the tower was keeping a close eye on the errant aircraft, because the next question was: "Cessna Zero Zero Zero Zero, are you landing on Runway Three?" To which the Cessna rogered again.

Now, Runways 01 and 03 happen to cross each other, and cross-traffic landings and takeoffs, if not controlled, can lead to messy situations. Fortunately for at least two aircraft, traffic was sufficiently spaced and the controller sufficiently alert to permit the visiting Cessna, which was now on a tight base, to continue to Runway 03. No harm was done.

Whether Confusion or Stupid was in command doesn't matter. We do know that it is unlikely that Approach vectored the pilot over and well north of the airport when a straight-in-approach from the south to Runway 01 was the shortest distance between the two points. We do know that the pilot must have confused Runway 03 with Runway 01, and we do know that he was flying a downwind at 210 degrees instead of the required 190 degrees. And we do know that he calmly rogered the question, "Are you landing on Runway Three?"

If the Cessna pilot had rogered less and repeated at least a couple of the instructions just once, the tower could have straightened things out before a potential hazard had arisen. Effective communication is *to share, to make common.* "Roger" doesn't necessarily ensure commonality of understanding.

Let's continue with the approach and landing. If you're transponder equipped, you may receive no further instructions from the tower other than "Cherokee Six One Tango, cleared to land." To which you reply what? "Roger, cleared to land, Cherokee Six One Tango." Roger. You've got it.

On the other hand, and particularly if you have a transponder, the tower may be in frequent communication to advise you of other aircraft that are in your vicinity:

Twr:　　　*Cherokee Six One Tango, traffic is a Cessna at one o'clock, two miles, westbound at two thousand.*

[Pause while looking]

You:　　　Negative contact, Cherokee Six One Tango.

Or, if you spot the traffic: "Cherokee Six One Tango has the Cessna."

Another situation: you're advised of the Cessna at one o'clock. You don't see it, and so inform the tower. A minute or so later, you spot it at your two o'clock position. At this point, tell the tower—even though the traffic is well to your right and presents no possible hazard: "Cedar Tower, Cherokee Six One Tango has the Cessna." The tower will acknowledge your message, often with a "thank you."

In a similar scenario, you hear the tower call another aircraft in your general area, alerting the pilot to *your* presence: "Cessna Eight Niner Golf, traffic is a Cherokee at ten o'clock, two miles, also westbound. Altitude unknown."

You have a reasonably good idea that that's you, so get on the air and help everybody by reporting your altitude: "Cedar Tower, Cherokee Six One Tango is at two thousand eight hundred, descending to one thousand seven hundred (pattern altitude)." You may or may not get an acknowledgment, but that's beside the point. You've kept the other parties informed—at least the interested parties.

This is a good time to mention that when you're given traffic in a "clock" position (nine o'clock, one o'clock, etc.), the position is based on your *ground track*, as shown on radar. The traffic's position may be different from your point of view, because of your wind correction angle. For example, you're tracking over the ground at 270 degrees, but because of a northerly wind, your heading is 300 degrees. If you are advised that you have a "target" at "12 o'clock," that means that the other aircraft is on your 270 degree track, not straight ahead of the nose of your airplane at 300 degrees. Using this example, the target is actually at about 11 o'clock in relation to the direction in which your airplane is pointed. Keep this in mind, especially under strong wind conditions when you need to crab to maintain your desired course.

Now let's continue and say you're nearing the airport. Once again, the tower calls you:

> **Twr:** *Cherokee Six One Tango, you're number two to land behind the Mooney on downwind.*
>
> **You:** Roger, number two to land. Negative contact on the Mooney [or "and we have the Mooney"]. Cherokee Six One Tango.

When the spacing is proper between you and the Mooney, or the Mooney is about to touch down, the tower will give you final clearance:

> **Twr:** *Cherokee Six One Tango, cleared to land, Runway Three Six.*
>
> **You:** Roger, cleared to land Runway Three Six, Cherokee Six One Tango.

Once you're on the ground, be sure to stay on the tower frequency until you have turned off the active and the tower has cleared you to the ramp or advised you to contact Ground Control. There's no way of knowing what might be happening behind you that would require the tower to issue you an emergency instruction. The call from the tower will probably be brief—no more than "Cherokee Six One Tango, contact Ground point niner." Remember that the Ground Control frequency is *typically* 121.7, 121.8, or 121.9. Consequently, the first three digits are often dropped.

Occasionally, especially if the tower is busy, the controller may fail to tell you to contact Ground, even though you are clear of the runway. In such cases, go past the taxiway hold line, come to a complete stop, and call the tower:

> **You:** Tower, Cherokee Six One Tango going to Ground.
>
> **Twr:** *Cherokee Six One Tango, roger. Contact Ground.*

This clears you to leave the tower frequency. Remember, stay on the tower frequency until a change is authorized.

OTHER TRAFFIC PATTERN COMMUNICATIONS

Because of spacing, the tower wants greater separation between you and the aircraft ahead of you:

> **Twr:** *Cherokee Six One Tango, extend your downwind for spacing.*
>
> **You:** Roger. Cherokee Six One Tango. Will you call my base? [Meaning, "Will you tell me when I can turn to the base leg?"]

Twr: *Cherokee Six One Tango, affirmative.*

Twr: *Cherokee Six One Tango, turn to base. You're number two to land behind the Aero Commander on final.*

You: Roger, and we have the Commander [*or* "negative contact on the Commander"], Cherokee Six One Tango.

You're on final, 100 feet above touchdown, and an unauthorized aircraft or ground vehicle ventures onto the runway. To avoid a confrontation, the tower issues a command:

Twr: *Cherokee Six One Tango, go around!*

You: [No response is necessary. Pour on the coals and initiate the go-around procedure. Don't argue; don't debate. Just do what you're told *now*.]

You've been shooting touch-and-gos but would now like the option of making a touch-and-go, stop-and-go, or a full-stop landing. Make the request on the downwind leg so the tower can say yea or nay, based on the existing traffic:

You: Cedar Tower, Cherokee Six One Tango requests the option.

Twr: *Cherokee Six One Tango, cleared for the option.*

The option approach is especially useful during flight instruction to maintain an element of surprise for the student, because go-arounds and missed instrument approaches are also permitted as options.

On final approach, decide (or have your instructor decide) how you'll end the approach. No need to tell the tower, you're cleared for whatever you decide.

When you've had enough for the day, advise the tower of your intentions—again on the downwind leg:

You: Cedar Tower, Cherokee Six One Tango will be full stop this time.

Twr: *Cherokee Six One Tango cleared to land.*

You: Cleared to land, Cherokee Six One Tango.

At this point, the tower makes a request of you:

Twr: *Cherokee Six One Tango, can you land short and turn off on Runway Two One?*

You: Affirmative, Cherokee Six One Tango. [Assuming you can comply with the request.]

Don't just come back with a "roger." Can you or can't you? The response is either "Affirmative" or "Negative."

WHEN THERE'S NO GROUND CONTROL OR ATIS ON THE FIELD

During periods of light traffic, the tower controller may double as the ground controller. In this case, the tower will provide the taxi clearance and instructions. However, you do need to secure taxi approval from the tower after clearing the runway on landing, just as if Ground Control existed:

> **You:** Tower, Cherokee Six One Tango clear of the active. Going to Ground.
>
> *Twr:* *Cherokee Six One Tango, taxi to the terminal. Caution. Men and equipment on right side of taxi strip.*
>
> **You:** Roger. Have them in sight. Cherokee Six One Tango.

On departure, call Ground on the published Ground Control frequency. But don't be surprised if the tower controller answers. If there is no ATIS on the field, the controller will normally communicate most of the ATIS information. (If he doesn't, ask for it.) To illustrate, you're at the ramp, ready to taxi:

> **You:** Chestnut Ground, Cherokee One Four Six One Tango at the terminal. Ready to taxi, VFR Centerville. Request the numbers.
>
> *GC:* *Cherokee Six One Tango, Chestnut Ground. Taxi to Runway One Niner. Wind two two zero at one zero, altimeter three zero one five.*
>
> **You:** Roger, taxi to One Niner. Cherokee Six One Tango.

When landing, you have presumably made your initial call 15 miles or so out, so you know the active runway. The tower, however, may or may not have given you the winds and altimeter. If it hasn't, and if monitoring the tower's communications with other aircraft hasn't revealed the data, it's perfectly in order to request it: "Tower, Cherokee Six One Tango. Request winds and altimeter." If the winds are reported variable in direction and/or velocity, you may want an update just before landing.

> **You:** Tower, Cherokee Six One Tango, wind check.
>
> *Twr:* *Cherokee Six One Tango, wind two zero zero at eight, gusting to two zero.*
>
> **You:** Roger. Cherokee Six One Tango.

OBTAINING SPECIAL VFR CLEARANCE

Back in the Flight Service chapter, we discussed obtaining Special VFR clearance when operating within a Control Zone under less than VFR conditions with no tower on the field. Now let's assume that the same conditions exist at an airport which has a tower. The procedures and calls are basically the same as those when a Flight Service Station is involved.

First, check the ATIS and then contact Ground Control and request the SVFR:

You: Municipal Ground, Cherokee One Four Six One Tango at Jet Air with Information Kilo. Request Special VFR southbound.

GC: *Cherokee One Four Six One Tango, taxi to Runway Three Three. Clearance on request.* [This means that Ground is requesting clearance for you from the Air Route Traffic Control Center. It does *not* mean that the clearance is available to you on your request.]

You: Roger. Taxi to Three Three, Cherokee One Four Six One Tango.

GC: [In a few minutes] *Cherokee One Four Six One Tango, advise when ready to copy.*

You: [Assuming you are free to copy] Cherokee One Four Six One Tango ready to copy.

GC: *ATC clears Cherokee One Four Six One Tango to exit the Municipal Control Zone to the south. Maintain Special VFR conditions at or below two thousand while in the Control Zone. Report leaving the Control Zone.*

You: Understand Cherokee One Four Six One Tango cleared Special VFR to depart south, maintain two thousand or below while in the Zone, and report clear of the Zone.

GC: *Cherokee Six One Tango, readback correct. Contact Tower.*

You: Roger, Cherokee Six One Tango.

When you've finished the pre-takeoff check and are ready to go, switch to the tower, which will already be aware of your clearance. Thus the following:

You: Municipal Tower, Cherokee One Four Six One Tango ready for takeoff. Special VFR southbound.

Twr: *Cherokee Six One Tango, roger. Cleared for takeoff. Report when clear of the Zone.*

You: Wilco, Cherokee Six One Tango.

This time, instead of departing, you want to land at Municipal. Through ATIS or any other source, you find that the field is below VFR limits but adequate for a Special VFR. *Before* entering the Control Zone, call the tower:

You: Municipal Tower, Cherokee One Four Six One Tango is over Deep Lake, level at three thousand, squawking one two zero zero with Information Xray. Request Special VFR for landing Municipal.

> *Twr:*   *Cherokee One Four Six One Tango, roger. Remain clear of the Control Zone until further advised.*

Now circle, slow down, or do whatever is necessary to remain outside the Zone until you hear from the tower again:

> *Twr:*   *Cherokee Six One Tango, Municipal Tower. Clearance when ready to copy.*

> You: Cherokee Six One Tango ready to copy.

> *Twr:*   *Cherokee One Four Six One Tango is cleared to enter the Control Zone east of Municipal. Maintain Special VFR conditions at or below two thousand five hundred. Squawk one two zero five. Report right base for Runway Three Five.*

> You: Roger. Cherokee One Four Six One Tango cleared to enter the Control Zone east of Municipal. Maintain Special VFR conditions at or below two thousand five hundred. One two zero five and report right base for Three Five.

> *Twr:*   *Cherokee Six One Tango, readback correct.*

> You: Tower, Cherokee Six One Tango turning right base for Three Five.

> *Twr:*   *Cherokee Six One Tango, roger. Cleared to land, Runway Three Five. Winds three two zero degrees at five.*

Of course, if the airport offers Approach Control, contact Approach for this clearance instead of Tower. You will be instructed when to contact Tower for landing clearance.

CONCLUSION

There's nothing complicated about calls to and from the tower, however, some basic principles and practices do apply.

If you don't understand an instruction, ask the tower to repeat with:

- "Say again."

- "Request further instructions. Am not familiar with the area."

- "Your transmission was garbled. Please say again."

- "You were cut out. Say again."

- "Please speak more slowly."

- "I do not understand that instruction."

Don't keep the controller in the dark as to what you intend to do. Tell him *before* you embark on an action or a maneuver that might affect his ability to control the other traffic safely.

Don't depend on "roger" to convey your understanding. Except for simple or routine directions, repeat the basic instructions the controller has issued. But keep your repeats short and to the point. Don't clutter up the air with "*ers*" and "*ahs*" and unnecessary verbiage. Communicate only the guts of the instruction or question—and do so tersely but distinctly:

>✈ "Roger, cleared to land."

>✈ "Can make short approach."

>✈ "Cherokee Six One Tango holding short."

>✈ "Cherokee Six One Tango extending downwind."

If you're not sure, *ask*. This is a little different from not understanding. You may have understood the instruction but certain uncertainties linger in your mind:

✈ "Is Cherokee Six One Tango cleared for an east departure?"

✈ "Will you call my base?" (*You've been asked to extend your downwind.*)

✈ "Is Cherokee Six One Tango cleared to turn base?" (*You've been asked to extend your downwind, but the tower hasn't cleared you to turn to base—and the runway is disappearing behind you.*)

✈ "How do you read me?" (*You have doubts about the strength or clarity of your transmission.*)

✈ "Is Cherokee Six One Tango cleared to land?" (*You've followed all instructions, are on the final, but clearance to land has not been received. Don't land without it!*)

If the controller asks you a question that offers you an option, such as "Can you make a tight base?," don't merely come back with "Roger." Confirm that you can or can't do as requested: "Roger, Cherokee Six One Tango will make tight base." Or, "Affirmative on the tight base, Cherokee Six One Tango."

The same type of response is essential when the controller says "hold short," "would you prefer Runway Three to Three Six?," or "will this be touch-and-go or full stop?" The responses should be:

✈ Cherokee Six One Tango holding short. (*If you're going, your moving aircraft will communicate your intentions.*)

✈ Six One Tango will take Runway Three.

✈ Another touch-and-go, Cherokee Six One Tango.

Rely on "affirmative" or "negative," when applicable:

- ✈ Negative contact. Cherokee Six One Tango.

- ✈ Negative, base would be too tight. Cherokee Six One Tango. *(When asked if you could land on a different runway while in the traffic pattern.)*

- ✈ Affirmative. Over the water tower. Cherokee Six One Tango. *(When queried if this is your present position.)*

Keep the tower informed. You're told that you're number three to land behind a Baron. You see the Baron and advise the tower accordingly but then lose it in the haze or sun: "Tower, Cherokee Six One Tango has lost the Baron."

As we've indicated enough times, you and everyone else in the airport vicinity make up a team. The tower calls the plays and it expects compliance. If a certain "play" won't work, don't go off on your own. Tell the "coach" and get permission *before* embarking on some imaginative digression. Remember that you're in an Airport Traffic Area—which is a *controlled* area in the very literal sense of the word. When a given instruction would place you or your flying machine in potential jeopardy or cause you to violate a regulation, you have both the right and responsibility to so advise the tower. Even then, unless it is a dire emergency, *request* a deviation, and give your reason for the request.

When all parties function as a moment-in-time team, traffic flows as it should. And when it flows as it should, the folks in the tower can be the nicest people in the world. Let's help them help us.

9

TCAs, ARSAs, TRSAs, Etc.

The Cerritos, California, midair collision in 1986 accelerated or instigated a string of FAA regulatory changes designed to prevent a recurrence of such incidents. The changes basically affected Terminal Control Areas (TCAs) and Airport Radar Service Areas (ARSAs).

The most recent changes (at this writing) instituted Mode C transponder requirements in the vicinity of TCAs, ARSAs, and certain other multiclass airports, eliminated the TCA designatures (i.e. Group I,II,III), and added instruction requirements for student pilots wishing to operate solo in TCAs.

Many additional recommendations are still under consideration, including reclassification of all airspaces to meet International Civil Aviation Organization recommendations (e.g. Class A,B, or C airspace). How many will be adopted, and when, is open to question. In the meantime, you can expect to see more TCAs established at airports currently having only ARSAs and more ARSAs established at airports currently having only Terminal Radar Service Areas (TRSAs).

The radio communication procedures required to use any of the three radar services remain basic. Thus, the examples we cite in Chapter 10 should be valid regardless of future TCA revisions and phase-out of TRSAs. Because several TRSAs still exist around the country, however, we'll continue to refer to them, describe them, and give examples of the radio phraseology associated with TRSA "Stage III" service.

What follows are brief descriptions of the various terminal radar services, beginning with the most stringently controlled terminal airspace, the TCA. In the next chapter, we'll review the radio phraseology for each.

****

Assume that you want to go into an airport that has a TCA, an ARSA, a TRSA, or one that has none of those but does provide radar Approach and Departure Control. Or maybe even an airport that offers *non-radar* Approach/Departure Control service. What is mandatory or optional in each situation? What do you say, and when do you say it?

If you're at all typical of the average non-instrument-rated pilot, you might be a bit confused as to your responsibility. (Perhaps *uncertain* is a better word.) Confusion and/or uncertainty has kept many pilots from venturing into controlled traffic areas. Instead, they have opted for the smaller, less-convenient airports that pose no challenge to their communication expertise. TCAs, ARSAs, and TRSAs are a bit too much—a bit too threatening.

It's the fear of the unknown, the feeling of insecurity, that causes pilots to detour miles around a TCA or put down in some remote airfield far from their ground destination. Insecurity is a natural emotion when the unknown confronts us, but we believe that insecurity can be conquered through education and practice. In this chapter, and the ones that follow, we'll give our hypothesis a try.

TERMINAL CONTROL AREAS (TCAs)

The traditional description of a TCA is that of an upside-down wedding cake, each layer having a bottom (a floor) and all levels having a common ceiling. The core always surrounds the primary airport within the TCA, with the higher levels fanning out from the core.

FIGURE 9-1 is a generalized side view of a TCA. A top view is shown in FIG. 9-2. These are simplistic sketches, because TCAs are not uniform in size, shape, or floor/ceiling altitudes. Nevertheless, the sketches reflect the fundamental concept.

TCAs are located at the largest and busiest airports in the county, and include:

Atlanta*	Houston	Philadelphia
Boston*	Kansas City	Pittsburgh
Chicago*	Las Vegas	St. Louis
Cleveland	Los Angeles*	San Diego
Dallas*	Miami*	San Francisco*
Denver	Minneapolis	Seattle
Detroit	New Orleans	Washington, D.C.* (Andrews AFB and National)
Honolulu		New York* (Kennedy, LaGuardia, Newark)

TCAs are also planned for Baltimore, Charlotte, Houston-Hobby, Memphis, Orlando, Phoenix, Salt Lake City, Tampa, and Washington-Dulles.

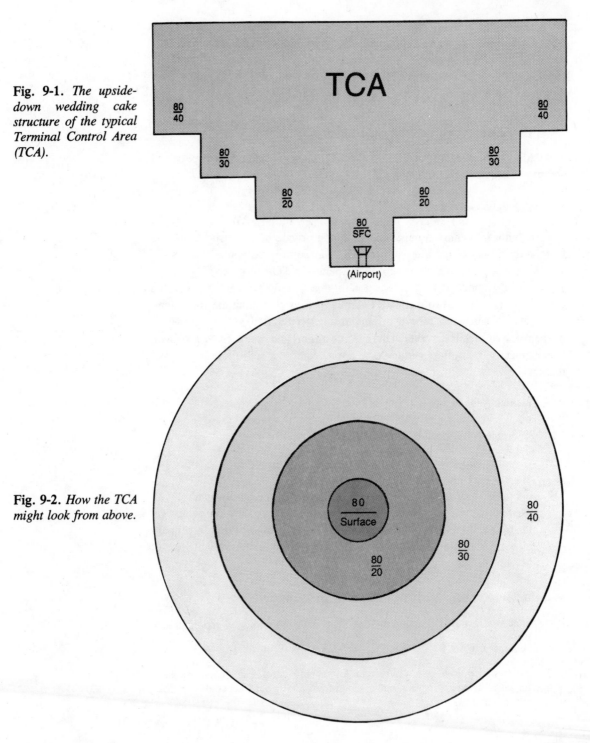

Fig. 9-1. *The upside-down wedding cake structure of the typical Terminal Control Area (TCA).*

Fig. 9-2. *How the TCA might look from above.*

The asterisks(*) indicate locations with special pilot requirements, discussed below. But regardless of these requirements, whether your destination is Atlanta or Seattle, before conducting *any* flight within the TCA—landing, taking off, or transiting—you must receive permission to do so. Without that, don't you dare set a prop blade into any part of it.

Pilot and Avionics Requirements

FARs 91.24 (91.215) and 91.90 (91.131) clearly establish both pilot and equipment requirements to enter a TCA:

Pilot Requirements

The pilot in command must hold at least a private pilot certificate to land or take off at any of the asterisked (*) TCA airports listed above.

A student pilot on a solo flight may operate in TCA airspace if he or she has received ground and flight instruction for that specific TCA and has received a logbook endorsement to that effect within 90 days preceding the solo flight.

A student pilot may takeoff or land at a non-asterisked TCA airport if he or she received such instruction *at* the specific airport and has received a logbook endorsement to that effect within 90 days preceding the solo flight to or from that airport.

Avionics Requirements

- operable VOR or TACAN receiver.

- operable two-way radio with frequencies appropriate for the TCA

- operable Mode 3/A 4096-code transponder (or Mode S transponder) with Mode C altitude reporting capability. Effective July 1, 1989 this equipment is also required in all airspace within 30 nautical miles of a TCA primary airport, from the surface up to and including 10,000 feet MSL. (See FIG. 9-3 and 9-5.)

Participation: Voluntary or Mandatory?

When entering or departing a TCA, you have no choice. Participation is *mandatory*. FAR 91.90 (91.131) makes this very clear:

No person may operate an aircraft within a terminal control area . . . unless that person has received appropriate authorization from ATC prior to operation of that aircraft in that area.

This should need no further elaboration. The dictum is ample warning to those who would intentionally or carelessly bust the TCA.

Identifying a TCA

TCAs are easy to identify on the Sectional because they are always depicted as solid blue concentric circles (although they often vary somewhat from true circles). Sectionals also highlight the existence of TCAs by heavy blue rectangles that extend for miles in all directions beyond the TCAs themselves. And TCAs are identified verbally on the chart in a rectangular box located just inside the heavy blue rectangles (FIG. 9-3).

And then, of course, there's the *Airport/Facility Directory*. Airports within a TCA are so indicated in the "Communications" section (FIG. 9-4).

Before flying into or near a TCA, be sure to equip yourself with the *current* VFR Terminal Area Chart (TAC) for the specific TCA. The TAC provides an enlarged and more detailed depiction of the TCA area, prominent landmarks, reporting points, Approach Control frequencies (based on your position), and other alerting and identifying features. It helps not only in navigating through the TCA, but also in avoiding the TCA altogether.

At this writing the FAA is developing a "TCA Avoidance Guide" for each TCA. The guides will include instructions and detailed large-scale maps to assist pilots in circumnavigating TCAs.

When and Whom to Call When Entering the TCA

We'll give examples of the calls themselves in the next chapter. Right now, let's merely establish the rules.

Using the Terminal Area Chart, *always* contact Approach Control on the designated frequency *before* you enter the TCA. FIGURE 9-5 will give you an idea of when you should call Approach. Of course, it doesn't hurt to call from farther out. On a busy day, you might be about to enter the TCA by the time Approach responds to your call.

Once Approach has approved your entry, you will be vectored through the TCA to a point at which Approach turns you over to the tower frequency for final landing instructions.

Let's set up another situation: The primary airport has sufficient traffic to warrant a TCA, but you want to land at another field that lies under the TCA but outside of the primary airport's immediate vicinity. St. Louis is a good example. The core of the TCA from the surface to 8000 feet surrounds Lambert Field. About 20 miles west-southwest is Spirit of St. Louis, a tower-controlled airport. From a side view, looking northward, the TCA appears as depicted in FIG. 9-6. You're approaching Spirit from the west. Are you required to contact

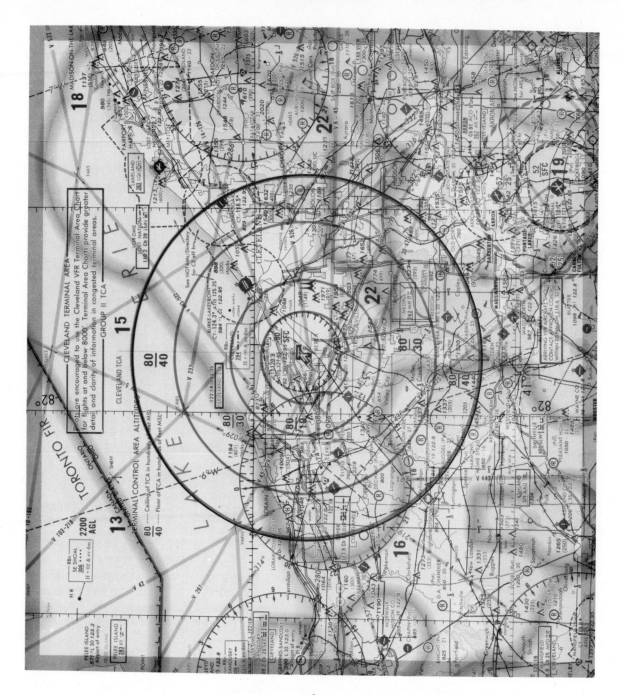

Fig. 9-3. *The big blue rectangle and solid blue concentric circles make TCAs unmistakable on the Sectional. We have added a circle to show the 60-mile-diameter area in which a Mode C transponder is required, effective July 1, 1989. Mode C will also be required in and above ARSAs, such as Akron-Canton at lower right, effective December 30, 1990.*

§   PHILADELPHIA INTL   (PHL)   5.2 SW   UTC–5(–4DT)   39°52'13"N 75°14'43"W   WASHINGTON
    21   B   S4   FUEL 100LL, JET A   OX 1, 2, 3, 4   LRA   CFR Index D   H-3C, L-24G, 28F
    RWY 09R-27L: H10499X200 (ASPH-GRVD)   S-100, D-350   HIRL CL   IAP
      RWY 09R: ALSF2. TDZ. Lights.   RWY 27L: MALSR. Ship. Tower.
    RWY 09L-27R: H9500X150 (ASPH-GRVD)   S-100, DT-350   HIRL CL
      RWY 09L: REIL. VASI(V6L)—Upper GA 3.25°TCH 93'. Lower GA 3.0°TCH 50.53'.
      RWY 27R: MALSR. Ship.
    RWY 17-35: H5460X150 (ASPH-GRVD)   S-75, DT-240   HIRL
      RWY 17: MALSR. VASI(V4L)—GA 3.0°TCH 54'. Tower.
      RWY 35: REIL. VASI(V4L)—GA 3.0°TCH 39'. Ship.
    AIRPORT REMARKS: Attended continuously. CAUTION advised for power back ops from airline gates—monitor ground
      control. Arpt management prohibits acft to make left turns from Taxiways CC, EE, GG, to Taxiway AA and from
      Taxiway V to Taxiway A and right turn from Taxiway X to Taxiway A. Taxiway Z unlgted, tfc restricted to acft less than
      25,000 pounds maximum gross ldg weight. Landing fee. Airport is located in a noise sensitive area. Company
      approved noise abatement takeoff procedures are to be used. 320' lgtd crane 3 NM east arpt indefinitely. Cargo
      ramp taxiway 'Q' to taxiway 'FF' avbl to cargo unit number 1 users only. Flight Notification Service
      (ADCUS) available.
    WEATHER DATA SOURCES: LLWAS.
    COMMUNICATIONS: ATIS 133.4   UNICOM 122.95
      PHILADELPHIA FSS (PNE) DL NOTAM FILE PHL.
    ® APP CON 126.6 (090°-269°) 125.4 (Rwy 17-35) 128.4 (270°-089°) 127.35
      TOWER 118.5   GND CON 121.9   CLNC DEL 118.85   PRE-TAXI CLNC 118.85
    ® DEP CON 119.75 (090°-269°) 124.35 (270°-089°)
      TCA Group: See VFR Terminal Area chart.
    RADIO AIDS TO NAVIGATION: NOTAM FILE PHL.
      DUPONT (L) VORTAC 114.0   DQO   Chan 87   39°40'41"N 75°36'27"W   066°21.1 NM to fld. 71/10W.
        NOTAM FILE ILG.
      JOANI NDB (LOM) 222   PD   39°54'04"N 75°05'43"W   263° 7 NM to fld.
      ILS 108.75 I-MYY Rwy 17
      ILS/DME 109.3 I-GLC Rwy 27L
      ILS/DME 109.3 I-PHL Rwy 09R
      ILS/DME 109.3 I-PDP Rwy 27R LOM JOANI NDB. BC unusable beyond 15 NM.
      ASR

Fig. 9-4. *A TCA identification in the* Airport/Facility Directory. *Use Clearance Delivery to obtain clearance to depart from a TCA airport.*

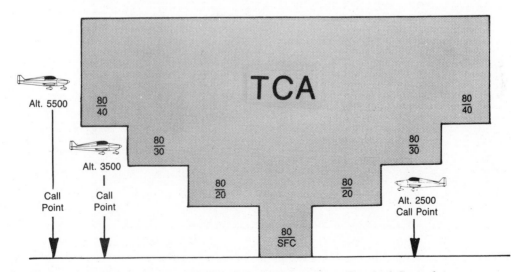

Fig. 9-5. *When to contact Approach Control if you want to enter a Terminal Control Area.*

Approach? *Yes,* if your position and altitude would put you into any one of the floor/ceiling levels. *No,* if you stay below the floors. You must, however, obtain clearance from Spirit's control tower before you enter its ATA.

So much for entering and landing in a TCA. How about taking off in a TCA? Here are four situations.

In the first case you are departing the primary TCA airport. You *must* receive a clearance from *Clearance Delivery* and, according to the instructions given you, you *must* contact Departure Control after takeoff.

All airports in TCAs and ARSAs, and some of those in TRSAs, have Clearance Delivery. Operating on its own frequency, its purpose is to communicate clearances in order to reduce congestion on the Ground Control frequency. Its availability at a given airport is identified in the "Communications" section of *Airport/Facility Directory* (FIG. 9-4).

In the second case you are departing a secondary airport under—not in—the TCA, such as Spirit of St. Louis. You want to go east and cruise at 5500 feet. As this altitude will quickly place you above the TCA floors, you *must* contact Approach Control before penetrating any floor.

Why Approach and not Departure Control? Good question. Leaving a secondary field you are going to *enter* the TCA. Therefore, you are *approaching* the TCA and thus under the control and surveillance of Approach. If you're departing from the primary airport, such as Lambert Field, you're already in the

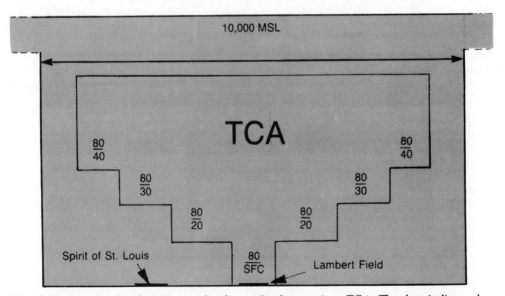

Fig. 9-6. *An example of an airport that lies* under *but not in a TCA. The dots indicate the Mode C requirement area effective July 1, 1989. This 30 nm radius from the primary airport clearly extends beyond the bounds of the TCA itself and encompasses outlying airports.*

core of the TCA and are departing from it, so Departure is the logical controlling agency.

In the third case you're taking off from a secondary airport such as Spirit and plan to fly locally or leave the area by staying well under the 2000, 3000, and 4000 foot floors. There is no requirement to call Approach because you're beneath the TCA, not *in* it. In this case, you merely establish the normal contact with Spirit Tower and stay tuned to the tower frequency at least until you're outside the five-mile ATA.

One more situation: you're on a cross-country and your route takes you directly through a TCA. You don't want to land at any airport within the area, however, and to go around the TCA would only add time and cost to the flight. If you're not sure of your radio procedures, you'll probably circumvent the whole thing and watch your fuel dwindle as the bill rises. TCAs can be transited, but only with permission, and only if you meet the pilot and equipment requirements.

If you're headed east, say from Kansas City to Cincinnati, at 7500 feet, the direct route is over St. Louis. Being cost-conscious, you call St. Louis Approach 30 miles or so out, identify yourself, give your position, altitude, destination, squawk, and request vectors through the TCA. Approach will take it from there and notify you when radar service is terminated once you're clear of the TCA on the east side.

Naturally, if you're above the ceiling—say at 9500 feet over St. Louis—or below the floors, no contact is needed. Otherwise, entry permission is required.

Remember though, effective July 1, 1989, you must be Mode C equipped to overfly a TCA, to fly beneath its floors, or to get anywhere within 30 nautical miles of the primary airport.

AIRPORT RADAR SERVICE AREAS (ARSAs)

In between the regulatory controls of the TCA and the TRSA comes the Airport Radar Service Area, the ARSA. Less restrictive than a TCA but more so than a TRSA, ARSAs are not only replacing TRSAs but are being established at airports that previously had neither.

To qualify for ARSA consideration, the FAA has stipulated that the primary airport must have a control tower, be served by a radar Approach/Departure Control facility, and meet at least one of the following conditions:

1. A minimum of 250,000 passenger enplanements a year, *or*

2. At least 75,000 instrument operations a year, *or*

3. At least 100,000 instrument operations a year at the primary and secondary airports that are included in the ARSA.

Identifying the ARSA And Its Structure

You can spot an ARSA on the Sectional by two thick slashed magenta concentric circles surrounding the primary airport (FIG. 9-7). Like a TCA, the exact shape may vary slightly. The *Airport/Facility Directory* also identifies ARSAs (FIG. 9-8).

Structurally, the ARSA is simple. It has only two rings, or circles, with the inner circle rising approximately 4000 feet above the primary airport. The radius of the inner circle extends five nautical miles from the airport. The second, or outer, circle has a 10 nautical mile radius, with a 1200-foot AGL floor and the same approximate 4000-foot AGL ceiling.

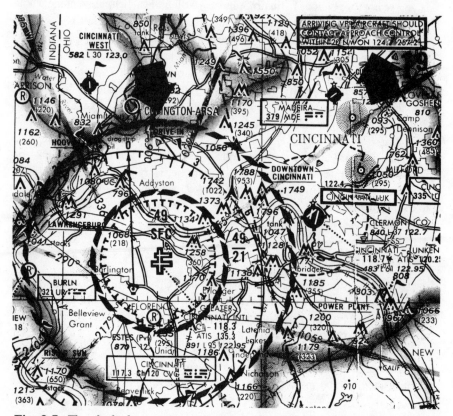

Fig. 9-7. *The slashed magenta concentric circles that identify an ARSA on the Sectional. The heavy rectangle at upper right advises of the frequency to be used by arriving VFR aircraft.*

CINCINNATI/COVINGTON, KY

§ GREATER CINCINNATI INTL   (CVG)   8 SW   UTC-5(-4DT)   39°02'52"N 84°40'00"W   CINCINNATI
891   B   S4   FUEL 100LL, JET A   LRA   CFR Index D   H-4H, L-21D, L-22E
RWY 18-36: H9501X150 (ASPH-GRVD)   S-75, D-185, DT-290, DDT-b80   HIRL CL   0.4% up N.   IAP
  RWY 18: SSALR. TDZ. VASI(V4R)—GA 3.0° TCH 52'.   RWY 36: ALSF2. TDZ.
RWY 09R-27L: H7800X150 (CONC)   S-75, D-185, DT-290, DDT-680   HIRL CL
  RWY 09R: MALSR. Tree.   RWY 27L: MALSR. VASI(V4L)—GA 3.0° TCH 60'. Tree.
RWY 09L-27R: H5500X150 (ASPH)   S-75, D-98   MIRL
  RWY 09L: Tree.   RWY 27R: VASI(V4L)—GA 3.0° TCH 37'. Road.
AIRPORT REMARKS: Attended continuously. Flocks of birds on or near arpt dalgt hours. Rwys 09R-27L and 18-36 gross
  weight strength for DC-10 and L-1011 aircraft is 510,000 pounds. Flight Notification Service (ADCUS) avbl.
WEATHER DATA SOURCES: LLWAS.
COMMUNICATIONS: ATIS 135.3   UNICOM 122.95
  LOUISVILLE FSS (LOU) Toll free call, dial 1-800-WX-BRIEF. NOTAM FILE CVG.
Ⓡ CINCINNATI APP CON   119.7 Rwy 18-36 (180°-359°) Rwy 09R-27L (090°-269°)
  124.7 Rwy 18-36 (360°-179°) Rwy 09R-27L (270°-089°) 121.0
  CINCINNATI TOWER 118.3   GND CON 121.7   CLNC DEL 121.3
Ⓡ CINCINNATI DEP CON   119.7 Rwy 18-36 (180°-359°) Rwy09R-27L (090°-269°)
  124.7 Rwy 18-36 (360°-179°) Rwy 09R-27L (270°-089°) 121.0 123.875
ARSA ctc APP CON
RADIO AIDS TO NAVIGATION: NOTAM FILE CVG.
  CINCINNATI (L) VORTACW 117.3   CVG   Chan 120   39°00'57"N 84°42'13"W   043° 2.3 NM to fld. 880/00.
  BURLN NDB (MHW/LOM) 321   UR   39°02'45"N 84°46'23"W   092° 4.3 NM to fld.
  ADDYS NDB (LOM) 351   SI   39°07'30"N 84°40'10"W   182° 3.9 NM to fld.
  ILS 108.7 I-JDP Rwy 27L
  ILS 111.5 I-SIC Rwy 18 LOM ADDYS NDB
  ILS 111.9 I-URN RWY 09R LOM BURLN NDB
  ILS/DME 109.9   I-CVG   Chan 36   Rwy 36
  ASR

Fig. 9-8. *Determining an ARSA in the* Airport/Facility Directory.

(Note in FIG. 9-7 that the 5-NM radius ARSA inner circle is slightly larger than the 5-SM radius Control Zone [and the 5-SM radius Airport Traffic Area, uncharted].)

So far, the ARSA's shape is rather like a TCA, except that the ARSA is considerably smaller in radius than either a TRSA or a TCA. There is one design feature, however, that makes ARSA different; it's called the *outer area*. This begins at the edge of the outer circle and extends for another 10 miles, giving the ARSA, in effect, a 20-NM dimension. In addition, the outer area rises from the lower limits of radio/radar coverage up to an altitude of 10,000 to 12,000 feet AGL, or, as the *AIM* explains it, " . . . up to the ceiling of the Approach Control's delegated airspace." FIGURE 9-9 illustrates the typical horizontal configuration of an ARSA. Referring back to FIG. 9-7, though, you'll notice that the outer area is never depicted on a Sectional.

Pilot and Equipment Requirements

ARSAs have no special pilot requirements. The only equipment needed is a two-way radio capable of sending and receiving on the ARSA frequencies, and (effective December 30, 1990) a Mode C transponder for operations within an ARSA and above it, up to and including 10,000 feet MSL.

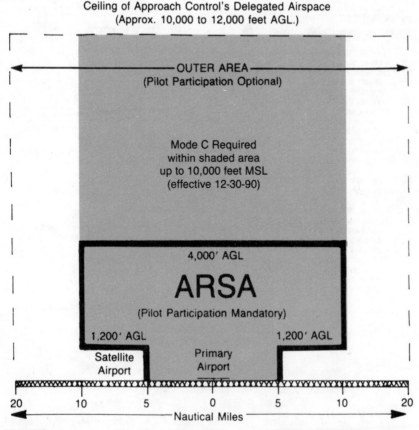

Fig. 9-9. *A cross-section of a typical ARSA. Actual ARSAs will vary from this generalized view.*

Pilot and Controller Responsibilities in the ARSA

From the pilot's point of view, and according to FAA regulations, no "clearance" into the ARSA is required. As FAR Part 91.88 states:

Arrivals and Overflights. No person may operate an aircraft in an airport radar service area unless two-way radio communication is established with ATC prior to entering that area and is thereafter maintained with ATC while within that area.

All the pilot has to do is contact Approach Control and establish radio contact to enter the ARSA. Unlike the TCA, you do not have to be literally "cleared" or have your entry into the ARSA "approved." (See additional comments on p. 116.)

The same regulation applies to departures, except that aircraft departing a satellite airport within the inner circle must establish radio communications as soon as practicable and then maintain radio contact while in the ARSA.

Once radio contact has been established, ATC will provide the following services within the 10 mile ARSA radius:

- Sequencing of all aircraft arriving at the primary airport.

- Maintaining standard separation of 1000 feet vertically and 3 miles horizontally between IFR aircraft.

- Traffic advisories and conflict resolution between IFR and known VFR aircraft so that the radar targets do not touch, or 500 feet vertical separation. (ATC will issue advisories or safety alerts if the targets are appearing to merge and the "green between" the targets on the scope is diminishing.)

- Issuing advisories and, if necessary, safety alerts to VFR aircraft.

The last point warrants clarification for those operating VFR. Unlike the TCA or TRSA, ATC does not *separate* VFR aircraft in the ARSA. Separation involves whatever vectoring or altitude changes are necessary to maintain a given distance between aircraft. *Advisories* alert the pilot to bearings, approximate distance, and altitude in relation to other aircraft which might pose a potential safety problem. It is then the pilot's responsibility to scan the skies, locate the traffic, and advise ATC when the traffic has been spotted.

A *safety alert* is issued when conflict between two aircraft, or an aircraft and a ground obstruction, seems imminent. When transmitted, it means that the receiving pilot should change course and/or altitude *now*. The situation is reaching emergency proportions. Alerts will be few, however, if advisories are heeded and the pilot is attentive to what's going on outside the cockpit.

The Outer Area

The outer area is not regulated airspace—not really part of the ARSA—therefore two-way radio communication in the outer area is not mandatory. Although controllers are learning to live with it, this outer area has been a source of some dispute. It can be at once a plus and a minus for pilots, while at the same time placing an added workload on the controller. To explain:

A pilot departs the primary airport and goes to the outer area for touch-and-gos at a satellite or to practice maneuvers. The controller *must* provide the same advisory and safety alert services if the pilot so requests, as if the pilot were operating within the ARSA *unless* the pilot requests termination of those services. It's up to the pilot to request or say "no thanks" when he's in the outer area; the controller has no such prerogative.

The potential—or real—problem lies in the fact that a whole lot of aircraft could be milling around in the outer area, which extends up to about 10,000 feet. ATC, having to provide the ARSA's service to these aircraft, could be so busy that it would be forced to temporarily deny landing or transiting aircraft entry into the ARSA.

The policy states that clearance into the ARSA is not required, but some controllers have found that they have had to tell pilots who are at the fringes of the outer circle (not the outer area) to "remain clear." The volume of activity in the outer area has, in some cases, created such an additional workload that service to those wanting to enter the ARSA has been delayed. Consequently, as things stand today, the policy can benefit one group and adversely affect another—with the controller caught in the middle.

As the *AIM* says, "While pilot participation in [the outer] area is strongly encouraged, it is not a VFR requirement." Not debating the value of radar service or your right to receive it, it would be considerate of others to terminate the service if and when you're going to fly in and around the outer area. If there's a valid reason for wanting the service, fine, but too many pilots who require advisories only consume air time, add work for the controller(s), and possibly delay others' entry *into* the ARSA.

So, let's summarize the ARSA concept this way:

- The only equipment required is a two-way radio capable of communicating with ATC and (beginning December 30, 1990) a Mode C transponder for operations within the ARSA and above it, up to and including 10,000 feet MSL.

- Radio contact with ATC *must* be established before entering the ARSA.

- ATC assumes that ARSA service is wanted as soon as radio contact is established. (If you call Approach and the controller responds to you with "Cherokee One Four Six One Tango, stand by," you have satisfied the ARSA communications requirements and may enter the ARSA, unless you are told to remain clear of the ARSA.)

- It is not clear whether ATC can legally refuse entry into the ARSA, since no "clearance" by ATC is required. In reality, though, controllers have been instructed that they *can* deny entry "when absolutely necessary" due to traffic saturation.

- Participating aircraft must comply with all ATC vectors, instructions, and other directions.

- ATC will provide standard separation between IFR aircraft (1000 feet vertically, three miles horizontally); minimum VFR-IFR separation (maintaining "green between" on the radar screen or 500 feet vertically); and advisories and, if necessary, safety alerts to VFR aircraft. No separation is provided between VFR aircraft.

- ATC will sequence landing aircraft in daisy-chain fashion and turn each aircraft over to the local control tower (primary or secondary airport) for final landing instructions.

✖ Outer area services are provided as long as radio communications are maintained. They are terminated only when the pilot advises that ARSA services are not desired.

✖ ARSA service to aircraft landing at satellite airports within the ARSA is discontinued when the pilot is instructed to contact that airport's control tower.

✖ Aircraft departing a satellite airport will not receive ARSA service until they have established radio communications and are radar identified.

✖ Participation is mandatory for any aircraft landing, departing, or transiting the 10-mile ARSA radius (except under the floor of the outer *circle*). It is voluntary if landing, departing, or operating within the outer *area*.

Conclusion

That is the story of ARSAs. The midair collision between a PSA 727 and Cessna 172 over San Diego in 1978 provided the impetus for a terminal control structure that would prevent another such incident. Then came the Cerritos, California, midair in 1986. This brought another burst of potential regulatory changes, including some major revisions of TCA requirements. Since then, several other incidents have occurred. These have triggered a flurry of new rules and proposals affecting traffic—particularly general aviation traffic—around the busier airports.

The ARSA, one result of these past experiences, had some start-up problems, and not a few pilots were—and are—critical of denied entrance into the ARSA, delays, "unnecessary" vectoring, and so on. Some of these problems, however, have been resolved as pilots and controllers alike gained experience in the new system. All in all, the ARSA is providing an additional measure of control around the busier airports where there had been little control before. That, in itself, is a plus—one more step toward greater safety in the skies.

TERMINAL RADAR SERVICE AREAS (TRSAs)

Now let's consider TRSAs. The major difference between TCAs, ARSAs, and TRSAs is that permission to enter, depart, or transit a TRSA is not mandatory. While the volume of flight activity may be considerable, a TRSA surrounds those airports that normally have less dense traffic. Harrisburg is not a Philadelphia; Binghamton is not a Kennedy.

Another difference: if you decide that you want radar vectors after entering a TRSA, you can call Approach Control at any time. As we've already emphasized several times, in a TCA and ARSA this is a no-no. Approach must be contacted *before* entering a TCA or ARSA.

Identifying a TRSA

While TRSAs are shaped much like TCAs, they can be quickly distinguished from TCAs on aeronautical charts because TRSAs are depicted by solid magenta lines. Also, as with a TCA, a TRSA identifying box is located just outside the outermost ring, as illustrated in FIG. 9-10.

In the *Airport/Facility Directory*, however, the only indication that a TRSA exists at a given airport is this notation under the COMMUNICATIONS section: "STAGE III SVC ctc APP CON" (or similar but appropriate instructions for the particular TRSA) (FIG. 9-11).

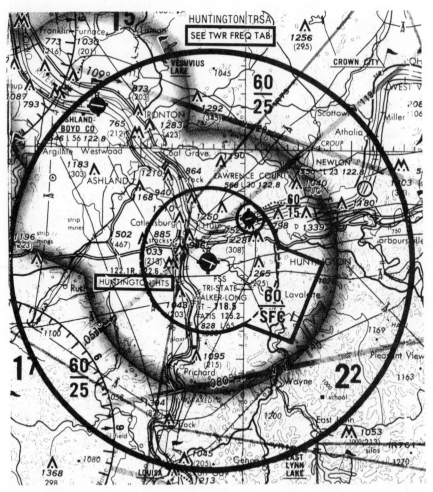

Fig. 9-10. *On the Sectional a TRSA resembles a TCA, except that the circles of a TRSA are magenta.*

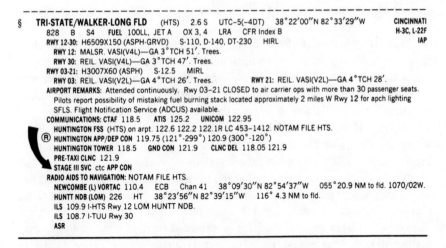

----

§   **TRI-STATE/WALKER-LONG FLD**   (HTS)   2.6 S   UTC−5(−4DT)   38°22′00″N 82°33′29″W   **CINCINNATI**
   828   B   S4   **FUEL** 100LL, JET A   OX 3, 4   LRA   CFR Index B   **H-3C, L-22F**
  RWY 12-30: H6509X150 (ASPH-GRVD)   S-110, D-140, DT-230   HIRL   **IAP**
    RWY 12: MALSR. VASI(V4L)—GA 3°TCH 51′. Trees.
    RWY 30: REIL. VASI(V4L)—GA 3°TCH 47′. Trees.
  RWY 03-21: H3007X60 (ASPH)   S-12.5   MIRL
    RWY 03: REIL. VASI(V2L)—GA 4°TCH 26′. Trees.   **RWY 21:** REIL. VASI(V2L)—GA 4°TCH 28′.
  AIRPORT REMARKS: Attended continuously. Rwy 03−21 CLOSED to air carrier ops with more than 30 passenger seats.
    Pilots report possibility of mistaking fuel burning stack located approximately 2 miles W Rwy 12 for apch lighting
    SFLS. Flight Notification Service (ADCUS) available.
  COMMUNICATIONS: CTAF 118.5   **ATIS** 125.2   **UNICOM** 122.95
    HUNTINGTON FSS (HTS) on arpt. 122.6 122.2 122.1R LC 453−1412. NOTAM FILE HTS.
  Ⓡ   **HUNTINGTON APP/DEP CON** 119.75 (121°-299°) 120.9 (300°-120°)
    HUNTINGTON TOWER 118.5   **GND CON** 121.9   **CLNC DEL** 118.05 121.9
    PRE-TAXI CLNC 121.9
    STAGE III SVC ctc **APP CON**
  RADIO AIDS TO NAVIGATION: NOTAM FILE HTS.
    NEWCOMBE (L) VORTAC 110.4   ECB   Chan 41   38°09′30″N 82°54′37″W   055°20.9 NM to fld. 1070/02W.
    HUNTT NDB (LOM) 226   HT   38°23′56″N 82°39′15″W   116° 4.3 NM to fld.
    ILS 109.9 I-HTS Rwy 12 LOM HUNTT NDB.
    ILS 108.7 I-TUU Rwy 30
    ASR

Fig. 9-11. *If the* Airport/Facility Directory *states "Stage III . . .", you know it's a TRSA.*

TRSAs are often more circular in shape than TCAs. They do, however, have the same basic upside-down wedding cake structure, with varying floor levels and a common ceiling. Thus, from a physical point of view, they closely resemble the TCAs.

Stage III Service

The service Approach and Departure Control provides in a TRSA is called *Stage III Service*. Originally there was a Stage I, Stage II, and Stage III, each offering more and more assistance to the pilot as the Roman numerals increased.

However, the Stage I designation has since been dropped, and Stage II is found only at a few airports around the country. All non-TRSA radar airports still offer certain radar services, but for all practical purposes only the Stage III designation remains, and Stage III is the service that the TRSA offers—if you ask for it.

While TRSAs will eventually be things of the past, there are still several dozen of them in existence around the country, so you should be familiar with their services and structure.

Stage III, through Approach and Departure Control, provides standard sequencing and separation of IFR and *participating* VFR aircraft. "Participating" means that the VFR pilot has requested the service. While a transponder is not required, if you have a transponder you would be particularly foolish *not* to utilize Stage III because you won't be turned right and left for initial radar identification, as non-transponder aircraft might be. Stage III is there to help you and to increase the overall safety in the TRSA. And, as is so often quoted in pilot manuals, "VFR participation is urged, but it is not mandatory."

When landing at the TRSA airport or transiting the TRSA on a cross-country, merely call Approach, identify yourself, your position, altitude, destination, squawk, and request Stage III service (FIG. 9-12). Approach will then give you a discrete transponder code and take it from there. All you have to do is follow directions until you're turned over to the tower or are clear of the TRSA.

Taking off from a TRSA airport, a VFR pilot who wants Stage III service would call Clearance Delivery (if it exists on the field-this is rare because most highly active TRSAs have been converted to ARSAs) and give his full identification, destination and/or route, and requested altitude. Clearance Delivery would then advise the pilot the heading he is to assume while in the TRSA, his altitude, the squawk code, and the Departure Control frequency. Once the clearance has been copied, the pilot is expected to *read it back* and receive confirmation that the readback was correct. (If there is no Clearance delivery on the field, merely contact Ground Control and request the Stage III service through Ground.)

Next the pilot tunes to Ground Control and advises Ground that he "has clearance" and is ready to taxi. The usual tower contact is, of course, established prior to takeoff; when airborne, the pilot switches to the Departure Control frequency *after* the tower approves the change. Departure then takes over and provides whatever vectoring or other instructions are necessary to speed the pilot on his way out of the TRSA.

TCA, ARSA, TRSA, AND SELECTED RADAR APPROACH CONTROL FREQUENCIES

COLUMBUS ARSA	124.2 267.9 (280°-099°)
	132.3 279.6 (100°-279°)
COVINGTON ARSA	119.7 257.2 RWY 18/36 (180°-359°) RWY 9R/27L (090°-269°)
	124.7 257.2 RWY 18/36 (360°-179°) RWY 9R/27L (270°-089°)
DAYTON ARSA	127.65 294.5 (360°-090°)
	118.85 327.1 (091°-180°)
	134.45 316.7 (181°-359°)
GREENSBORO ARSA	120.9 233.2 (250°-049°)
	118.5 284.6 (050°-107°)
	126.6 322.3 (108°-249°)
LEXINGTON ARSA	120.15 259.3 (040°-220°)
	120.75 298.9 (221°-039°)
ROANOKE ARSA	119.05 339.8 (360°-150°)
	126.9 339.8 (151°-359°)
BRISTOL TRSA	118.4 349.0 (047°-227°)
	125.5 317.5 (228°-046°)
CHARLESTON TRSA	119.2 259.1 (050°-229°)
	124.1 259.1 (230°-049°)
HUNTINGTON TRSA	119.75 257.8 (121°-299°)
	120.9 257.8 (300°-120°)
BLACKSTONE AAF PERKINSON RADAR	124.05 353.9
RICKENBACKER ANGB RADAR	132.3 279.6

Fig. 9-12. *Unlike TCAs and ARSAs, TRSA frequencies are not printed on the map part of the Sectional. They're in a special table on the Sectional, and also in the* A/FD.

What if you, the pilot, don't want the Stage III service after departure? Then simply bypass Clearance Delivery, call Ground Control, give them the standard pre-taxi information, and conclude with "Negative Stage III Service."

When landing at or taking off from a TRSA airport, Approach or Departure Control is *required* to give you Stage III service *if you request it*. When landing at or taking off from an airport that *underlies* a TRSA, Stage III service, if requested, is provided on a "workload permitting" basis.

RADAR APPROACH AND DEPARTURE CONTROL AIRPORTS

Some airports have no TCA, ARSA, or TRSA yet are equipped with radar Approach and Departure Control. This cannot, however, be determined from the Sectional. As just one example, take Clarksburg, West Virginia (FIG. 9-13).

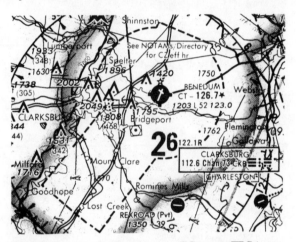

Fig. 9-13. *Clarksburg has no TCA, no TRSA, no ARSA, but it does offer radar Approach and Departure Control. The Sectional, however, won't tell you that.*

To all appearances, this is just a standard Control Zone airport with a tower. But, if you wanted to land at Clarksburg, a little research into the *Airport/Facility Directory* would reveal that radar Approach and Departure Control are on the field. How do you know? The symbol ® before "Approach" and "Departure" signifies the existence of radar (FIG. 9-14). The *A/FD* indicates that Stage II service is available, if you want it. Stage II includes advisories and sequencing, but *not* separation, of VFR aircraft.

Airports which do not have their own Approach and Departure Control facilities can often rely on the services of a neighboring radar airport (FIGS. 9-15 and 9-16) or an Air Route Traffic Control Center. These airports are discussed further in later chapters.

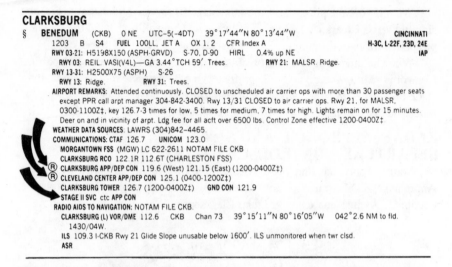

CLARKSBURG

§   **BENEDUM**   (CKB)   0 NE   UTC-5(-4DT)   39°17'44"N 80°13'44"W   **CINCINNATI**
   1203   B   S4   FUEL 100LL, JET A   OX 1, 2   CFR Index A   **H-3C, L-22F, 23D, 24E**
   RWY 03-21: H5198X150 (ASPH-GRVD)   S-70, D-90   HIRL   0.4% up NE   **IAP**
      RWY 03: REIL. VASI(V4L)—GA 3.44°TCH 59'. Trees.   RWY 21: MALSR. Ridge.
   RWY 13-31: H2500X75 (ASPH)   S-26
      RWY 13: Ridge.   RWY 31: Trees.
   AIRPORT REMARKS: Attended continuously. CLOSED to unscheduled air carrier ops with more than 30 passenger seats
      except PPR call arpt manager 304-842-3400. Rwy 13/31 CLOSED to air carrier ops. Rwy 21, for MALSR,
      0300-1100Z‡, key 126.7-3 times for low, 5 times for medium, 7 times for high. Lights remain on for 15 minutes.
      Deer on and in vicinity of arpt. Ldg fee for all acft over 6500 lbs. Control Zone effective 1200-0400Z‡
   WEATHER DATA SOURCES: LAWRS (304)842-4465.
   COMMUNICATIONS: CTAF 126.7   UNICOM 123.0
      MORGANTOWN FSS (MGW) LC 622-2611 NOTAM FILE CKB
      CLARKSBURG RCO 122.1R 112.6T (CHARLESTON FSS)
   ® CLARKSBURG APP/DEP CON 119.6 (West) 121.15 (East) (1200-0400Z‡)
   ® CLEVELAND CENTER APP/DEP CON 125.1 (0400-1200Z‡)
      CLARKSBURG TOWER 126.7 (1200-0400Z‡)   GND CON 121.9
      STAGE II SVC ctc APP CON
   RADIO AIDS TO NAVIGATION: NOTAM FILE CKB.
      CLARKSBURG (L) VOR/DME 112.6   CKB   Chan 73   39°15'11"N 80°16'05"W   042°2.6 NM to fld.
      1430/04W.
      ILS 109.3 I-CKB Rwy 21 Glide Slope unusable below 1600'. ILS unmonitored when twr clsd.
      ASR

Fig. 9-14. *Check the* Airport/Facility Directory. *Clarksburg offers Stage II radar service. When Clarksburg's own Approach/Departure Control is closed at night, Cleveland Center offers similar radar services.*

WILMINGTON

§   **GREATER WILMINGTON-NEW CASTLE CO**   (ILG)   4.3 S   UTC-5(-4DT)   **WASHINGTON**
   39°40'42"N 75°36'25"W   **H-3C, L-24G, 28F**
   80   B   S4   FUEL 100LL, JET A, MOGAS   OX 1, 2, 3, 4   LRA   CFR Index AA   **IAP**
   RWY 09-27: H7165X150 (ASPH-GRVD)   S-90, D-140, DT-250   HIRL
      RWY 09: ODALS.   RWY 27: VASI(V4L)—GA 3.0°TCH 50.7'. Tree.
   RWY 01-19: H7002X200 (ASPH-GRVD)   S-90, D-140, DT-250   HIRL
      RWY 01: ASLF1. Road.   RWY 19: VASI(V4L)—GA 3.0°TCH 58'. Tree.
   RWY 14-32: H5004X150 (ASPH)   S-50, D-60   MIRL
      RWY 14: Tree.   RWY 32: VASI(V4L)—GA 3.0°TCH 51.4'. Tree.
   AIRPORT REMARKS: Attended continuously. CAUTION: Flocks of birds on and in vicinity of arpt. CAUTION-240' cranes
      marked and lgtd within boundary of arpt 800-4000' west of Rwy 01-19 and 800-3900' north of Rwy 09-27
      indefinitely. CAUTION-2' by 6' panels located 20' left and 1500'; 1600' and 2000' from thld of Rwys 14 and 32.
      ACTIVATE ALSF1 Rwy 01 and ODALS Rwy 09 when twr clsd—126.0. Fee for all twin engine acft and acft over 7500
      pounds; no fee for single engine acft. Rwy 09-27 no touch and go ldg for turbo jet 0400-1200Z‡. When terminal
      building clsd 0400-1100Z‡ ctc arpt security on 121.7 or 302-573-3935. Overnight parking on terminal ramp
      restricted to acft over 12,500 pounds. Closed to unscheduled air carrier operations with more than 30 passenger
      seats except one hour PPR, call 302-323-2680. Rwy 14-32 CLOSED to FAR Part 121 operators except 1 hour PPR
      call 302-323-2680. Rwy 14-32 Clsd to acft over 12,500 lbs. except with prior approval.
   COMMUNICATIONS: CTAF 126.0   ATIS 123.95 (1200-0400Z‡)   UNICOM 122.95
      MILLVILLE FSS (MIV) Toll free call, dial 1-800-WX-BRIEF. NOTAM FILE ILG.
      DUPONT RCO 122.1R 114.0T (MILLVILLE FSS)
   ® PHILADELPHIA APP/DEP CON 125.0
      WILMINGTON TOWER 126.0 (1200-0400Z‡)   GND CON 121.7
   RADIO AIDS TO NAVIGATION: NOTAM FILE ILG.
      DUPONT (L) VORTAC 114.0   DQO   Chan 87   39°40'41"N 75°36'27"W   at fld. 71/10W.
      HADIN NDB (LOM) 248   IL   39°34'52"N 75°36'52"W   015° 5.3 NM to fld.
      ILS 110.3 I-ILG Rwy 01 LOM HADIN NDB. ILS unmonitored when twr clsd.

Fig. 9-15. *The Wilmington (Delaware) airport has its own tower, but uses Philadelphia's radar Approach and Departure Control.*

GAITHERSBURG

§ **MONTGOMERY CO AIRPARK** (GAI) 2.6 NE UTC–5(–4DT) 39°10′05″N 77°09′59″W WASHINGTON
540 B S4 FUEL 100LL, JET A OX 4 TPA—1340(800) L-24G, 28E, A
RWY 14-32: H4235X75 (ASPH) MIRL IAP
RWY 14: REIL. VASI(V4L). Thld dsplcd 200′. RWY 32: REIL.VASI(NSTD). Trees. Rgt tfc.
AIRPORT REMARKS: Attended 1300-0100Z‡. CAUTION–hazardous terrain and trees 50′ W Rwy 32. ACTIVATE MIRL Rwy
14–32 and REIL Rwy 32—122.85. REIL Rwy 14 out of svc indefinitely. RWY 32 Collins experimental VASI. Noise
abatement depart Rwy 32 turn rgt to at least 340°, refrain from Rwy 32 tkf between 0400–1200Z‡. Helicopter TPA
600 AGL.
COMMUNICATIONS: CTAF/UNICOM 122.7
LEESBURG FSS (DCA) Toll free call, dial 1–800–WX–BRIEF. NOTAM FILE DCA.
® BALTIMORE APP/DEP CON 128.7
BALTIMORE CLNC DEL 121.6
RADIO AIDS TO NAVIGATION: NOTAM FILE DCA.
ARMEL (L) VORTAC 113.5 AML Chan 82 38°56′04″N 77°28′01″W 053°19.7 NM to fld.
297/08W.NOTAM FILE IAD.
FREDERICK (T) VOR 109.0 FDK 39°24′44″N 77°22′32″W 155°17.3 NM to fld. NOTAM FILE MRB.
GAITHERSBURG NDB (MHW) 385 GAI 39°10′04″N 77°09′50″W at fld.

Fig. 9-16. The Gaithersburg (Maryland) airport is an uncontrolled field northwest of Washington, D.C. Yet its radar Approach and Departure Control services are provided by Baltimore's airport.

NON-RADAR APPROACH AND DEPARTURE CONTROL

The least sophisticated type of Approach and Departure Control is of a non-radar nature. Airports where this type of service is available can be determined only from the *Airport/Facility Directory*—not the Sectional. If no ® precedes "APP CON" or "DEP CON," then radar control does not exist.

So if there's no radar, what good is the service? Well, it's certainly a non-precision service, but it still can be useful. What happens is this: you call Approach 20 or so miles out and provide the usual information of identification, position, altitude, and destination. Since there's no radar, there's no point in quoting your squawk code. The controller acknowledges your call, tells you what altitude to fly, where to report, and when to contact the tower. Basically, the controller is tracking you by memory based on your position reports. Without the benefit of radar but with accurate position and altitude reporting by participating pilots, he or she can provide a reasonable separation and sequencing of aircraft in the area. Again, it's hardly precise, but it's better than nothing.

APPENDIX D

Airports/Locations Where the Transponder

Requirements of § 91.24(b)(5)(ii) Apply

Section 1. The requirements of § 91.24(b)(5)(II) apply to operations in the vicinity of each of the following airports:

Logan International Airport, Billings, MT.
Hector International Airport, Fargo, ND.

Fig. 9-17

DESIGNATED AIRPORT AREAS

Effective December 30, 1990, Mode C transponders are required from the surface to 10,000 feet MSL within a 10 nautical mile radius of any airport listed in Appendix D of FAR Part 91, excluding the airspace below 1,200 feet AGL outside of the Airport Traffic Area for that airport (FIG. 9-17). Initally, two airports are affected: Billings, Montana, and Fargo, North Dakota. Airports chosen to be included in Appendix D have high passenger traffic (200,000 enplanements annually), existing radar service, substantial instrument operations, yet are not planned for an ARSA.

	Pilot Participation Required	Two-Way Radio Communication Required	Mode C Transponder Required[4]	VOR Required	Special Pilot Requirements[8]	"Clearance" Required	Chart Indication
TCA	✓	✓	✓[6]	✓	✓	✓	Solid Blue Lines[7]
ARSA[1]	✓	✓	✓				Slashed Magenta Lines
Outer Area Near ARSA	3	3	5				Not Shown
Designated Airport Areas[2]	3	3	✓				Thin, Solid Dark Blue Lines
TRSA	3	3					Solid Magenta Lines
Other Radar Approach/Departure Control Airspace	3	3					Not Shown, Check A/FD
Non-Radar Approach/Departure Control Airspace	3	3					Not Shown, Check A/FD

[1]inner circle (including satellites) and outer circle. Services in outer area same as in outer circle, when requested.
[2]as listed in FAR 91 Appx. D.
[3]required if services are desired, otherwise required only within ATA.
[4]consult FAR 91.24 (91.215) for complete rules.
[5]includes airspace above ARSA up to 10,000 to 12,000' MSL.
[6]includes airspace within 30 nautical miles of primary TCA airport, up to
[7]lateral limits of Mode C requirement area are indicated by thin, solid, dark blue lines.
[8]see requirements earlier in this chapter.

Fig. 9-18. *Certain non-TCA, non-ARSA airports with high passenger traffic are designated to have Mode C transponder requirements, effective December 30, 1990.*

CONCLUSION

FIGURE 9-18 is a summary of the requirements of TCAs, ARSAs, TRSAs, radar and non-radar Approach/Department Control, and Designated Airport Areas. In only two cases is VFR pilot participation mandatory: when entering, departing, or transiting a TCA or ARSA. Those, in every respect, are strictly controlled areas. Going over, under, or around either requires no participation or approval, nor does flying (for whatever purpose) in a TRSA, in a Designated Airport Area, or in the vicinity of the less sophisticated radar and non-radar airports (except, of course, within an ATA). Mode C, however, may be required, depending on altitude and position. IFR requirements are much more stringent.

None of this says, however, that participation isn't wise. By all standards, it is. At the same time, operating in a radar-controlled environment does not relieve the VFR pilot of the responsibility for maintaining constant surveillance of the skies surrounding him. It's too easy to be lulled into a false complacency when someone on the ground is vectoring you, alerting you to real or possible threats, and generally directing you and others through the melange of aircraft—big and small.

If there's one failure we've noticed among professional instructors, students, and licensed pilots, it's their propensity to keep their eyes either in the cockpit or straight ahead. They apparently haven't been taught to develop a swivel neck and maintain a constant scanning of as much of a 360-degree circle as aircraft structure and physical limitations will permit. Back in the dark ages of World War II, this was a point of continual emphasis during pilot training. You could never be sure what unfriendly fellow was out there with murder in his heart. The best insurance was to see him first, take evasive action, and then fire away.

Today it isn't considered proper to fire at potential targets, but seeing and evading are more than acceptable practices; they are *lifesaving*. If you have your head in the cockpit, for whatever reason, more than 10 seconds in VFR conditions, you ought to be getting a little nervous. That's too long, considering the rate of closure between even light aircraft. Flying in instrument conditions is one thing; attention to the instruments is essential. VFR is something else, "Head up and unlocked" should be the byword of every pilot.

All of which has nothing to do with radio communications—except to observe that in the process of communicating and being monitored on radar, do not allow yourself relaxed alertness. Your safety in flight is enhanced by those on the ground who are directing and advising you, but you still have the responsibility to be your own protector and defender. A swivel neck and sharp eyes are the best shields against disaster in the air.

Let's go now to examples of the radio contacts between you and the various ground agencies we have been reviewing.

10

Departure and Approach Control

There are three primary communication and control agencies in the world of flight: Tower, Center, and Approach/Departure Control.

WHERE ARE DEPARTURE AND APPROACH LOCATED?

We mean by that question the physical location on the airport, not which airports provide the service.

To respond directly, both are in the control tower or in a radar-equipped facility that ensures the ability of both functions to coordinate closely with the tower personnel. Keep this fact in mind, however: In a TCA, ARSA, or TRSA, Approach and Departure are based on the primary field that warrants the TCA, ARSA, or TRSA designation. If you're landing at Spirit of St. Louis or Kansas City's Downtown Airport, you're not *in* but *under* the TCA. You'll thus be talking to an Approach or Departure controller who is physically located at Lambert Field or Kansas City International—some 15 to 20 miles away. That separation, however, in no way affects the controller's ability to coordinate instantly with the non-TCA control tower.

The tower is always contacted in a controlled airport and Center should be brought into the picture on cross-country VFR trips, as we'll explain in the next chapter. The purpose of Approach and Departure, then, is to serve as an intermediary between Tower and Center for departures, and Center and Tower for arrivals. In other words, Approach and Departure provide an extra hand to sequence and separate IFR and participating VFR aircraft to ensure a smooth flow of traffic in the TCA, ARSA, or TRSA. Without these middlemen, the towers at busy airports would face an impossible task of vectoring, sequencing, separating, and directing landing and departing aircraft. It just couldn't be done while still preserving a modicum of safety.

On arrival, Center (if it has been in the picture) hands off the inbound aircraft to Approach at a given point. Approach guides the pilot through the TCA, ARSA, or TRSA until he is about to enter the traffic patteren, and then turns him over to the tower frequency.

In a departure, the process is reversed: Tower to Departure to Center. Thus in a cross-country between two TCAs, the sequence from beginning to end is: ATIS—Clearance Delivery—Ground Control—FSS (to open the flight plan)—Tower—Departure—Center—ATIS—Approach—Tower—Ground Control—FSS (to close the flight plan).

This should not imply that Center *must* be used, but, as we'll discuss in the next chapter, why not? And, obviously, if the departure is from an airport that underlies a TCA or ARSA and no portion of the flight will be conducted in the TCA or ARSA, Departure does not have to be involved. So there are variations. The sequence cited, however, includes all possible traffic control agencies, omitting only the en route contacts with Flight Service. Flight Service, however, is in no respect a *controlling* agency.

DEPARTURE FROM A TCA AIRPORT

To become more specific about all of this, along with examples of the reccomended radio phraseology, let's say that you're on the ramp at Kansas City International, which as you know by now, is a TCA. The engine is started and you're ready to go.

First, tune to ATIS and remember the information as well as the phonetic designation of the report. Now dial in Clearance Delivery (or, preferably, Pre-Taxi Clearance, if available); the frequency is listed in the *A/FD*. Listen to be sure the air is clear and then announce yourself:

You:	International Clearance, Cherokee One Four Six One Tango. [Kansas City.]
CD/PTC:	*Cherokee One Four Six One Tango, Clearance.*
You:	Clearance, Cherokee One Four Six One Tango on the general aviation ramp, VFR to Omaha, request six thousand five hundred.
CD/PTC:	*Cherokee Six One Tango, roger. Cleared to depart the International TCA. Fly runway heading, maintain three thousand, expect six thousand five hundred, ten minutes after*

departure. Departure Control frequency one one niner point zero. Squawk one two five two.

You: Understand Cherokee One Four Six One Tango cleared to depart the TCA, runway heading, three thousand, six thousand five hundred in ten, one one niner point zero, one two five two.

CD/PTC: *Cherokee Six One Tango, readback correct. Contact Ground.*

You: Roger, will do. Cherokee Six One Tango.

As we said in the last chapter, when obtaining a VFR clearance at a TCA airport, you're expected to read back the instructions to ensure your understanding. If you're prepared for what's coming and write down the instructions, you'll save everyone a lot of time and yourself a lot of confusion. The normal delivery sequence is just as we have indicated: clearance to depart—initial heading—initial altitude—time to be at the desired altitude—Departure frequency—squawk code. So much of effective radio communications is knowing what to expect to hear and being ready to respond accordingly. This situation is no exception.

Now comes the call to Ground Control:

You: International Ground, Cherokee One Four Six One Tango at general aviation ramp with clearance and Information Charlie. Ready to taxi.

GC: *Cherokee One Four Six One Tango, taxi to Runway One Eight.*

Two points: When Clearance Delivery gave you your clearance, that was your permission to operate your aircraft in the TCA. That was it, and no further authority is required. Second, you were given a transponder squawk of 1252 in this case. Immediately set your transponder accordingly. Don't wait. You might forget either the code or to set it at all. And remember to turn the transponder switch to STANDBY—not the ON or ALT position. Change to ALT only after you've been cleared for takeoff and are taxiing from the hold line to the active runway.

The runup is complete and you're at the hold line. Now call the tower:

You: International Tower, Cherokee One Four Six One Tango ready for takeoff.

Twr: *Cherokee Six One Tango, roger. Fly runway heading. Cleared for takeoff.*

You: Roger. Cherokee Six One Tango. [Now turn your transponder to ALT.]

After you're airborne, the tower will contact you again:

Twr: *Cherokee Six One Tango, contact Departure.*

You: Roger. Cherokee Six One Tango. Good day.

Note that the tower said nothing about the frequency on which to contact Departure. Also, you didn't have to request the frequency change. The reason is that all factors relative to your clearance had already been coordinated between Clearance, Tower, and Departure.

If you have two radios, change the frequency in one from Ground Control to Departure just before calling the tower. If you can set up your radios so that you're always one step ahead of the facility you're currently talking to, the transition from one to the next will be simplified. We discussed this in the Control Tower chapter, but it's worth reemphasizing as just one more element of good cockpit organization.

The next call comes following the tower's approval to contact Departure:

You: Kansas City Departure, Cherokee One Four Six One Tango is with you, out of one thousand four hundred for three thousand.
[Departure is well aware of your clearance and your squawk code. Consequently, all you need to do is identify your aircraft, your present altitude, and the altitude to which you are climbing. "With you" simply establishes the contact.]

Dep: *Cherokee Six One Tango, climb and maintain six thousand five hundred. Turn right heading three four zero until receiving St. Joe, then direct.* ["Until Receiving St. Joe" means picking up the St. Joseph, Missouri, VOR, which is about 50 miles north of Kansas City. "Direct" logically means to head out on your own toward Omaha.]

You: Roger. Out of one thousand seven hundred for six thousand five hundred. Understand three five zero to St. Joe, then direct. Cherokee Six One Tango.

Dep: *Cherokee Six One Tango, that heading is three four zero to St. Joe. Otherwise correct.*

You: Roger. Three four zero on the heading. Cherokee Six One Tango.

A few minutes later:

Dep: *Cherokee Six One Tango, position two zero miles north of International, departing the Kansas City TCA. Stand by for traffic advisories.* [You can decline further radar service if you wish, but let's assume you don't.]

You: Roger. Cherokee Six One Tango. Standing by. [You may or may not hear anything further from Departure until you leave the radar coverage area as follows.]

Dep: *Cherokee Six One Tango, radar service terminated. Squawk one two zero zero, change to advisory frequency approved.* ["Advisory frequency" is any frequency you now wish—Center, Flight Service, en route airport towers, etc. There is a chance that radar service will not be terminated and that Departure will "hand you off" to the appropriate Air Route Traffic Control Center for further traffic advisories. We'll get into that in the next chapter.]

You: Roger. Cherokee Six One Tango. Good day. [Now change the transponder to 1200.]

DEPARTING AN AIRPORT UNDERLYING A TCA

You're departing a Spirit of St. Louis, a Kansas City Downtown, a Peachtree DeKalb (in Atlanta), or any other field that lies *under* but is not *in* a TCA. What differences, if any, does this situation pose as far as clearances, radio contacts, and the like are concerned?

Assume that you're departing Spirit and your route of flight is to the northeast. Your desired altitude will put you squarely in the TCA. There is no Clearance Delivery and no agency on the field that can authorize your penetration of the TCA, so the sequence of contacts goes like this:

You: Spirit Ground, Cherokee One Four Six One Tango at the terminal, ready to taxi, VFR northeastbound with Information Delta.

GC: *Cherokee One Four Six One Tango, taxi to Runway Two Five.*

The runup is complete, you've taxied to the hold line, the transponder is on 1200, the switch in STANDBY:

You: Spirit Tower, Cherokee One Four Six One Tango, ready for takeoff, northeast departure.

Twr: *Cherokee Six One Tango, cleared for takeoff. Northeast departure approved.*

You: Roger, cleared for takeoff, Cherokee Six One Tango.

As you're taxiing onto the runway, change the transponder switch to ALT and get ready to roll. When you are safely off the ground, the tower should say:

Twr: *Cherokee Six One Tango, contact St. Louis Approach, one two six point seven.*

You: Roger. One two six point seven, Cherokee Six One Tango.

Now, dial in 126.7—if you didn't know the frequency in advance—and call Approach:

You: St. Louis Approach, Cherokee One Four Six One Tango.

App: *Cherokee Six One Tango, St. Louis Approach.*

You: Approach, Cherokee Six One Tango just off Spirit, heading of zero three zero at one thousand eight hundred, requesting seven thousand five hundred to Decatur and clearance to transit the TCA.

App: *Cherokee Six One Tango, squawk zero two five six and ident. Remain outside the TCA until radar contact.*

You: Cherokee Six One Tango squawking zero two five six.

You must contact Approach Control at the primary TCA airport for clearance into the TCA. (Remember why it's Approach and not Departure? If your recollection is faint, go back to Chapter 9). If you do forget when the time comes and happen to call Departure instead of Approach, the controller won't subject you to public reprimands. Should you say "St. Louis Departure" and the controller acknowledges with "St. Louis Approach," go ahead with what you wanted to say and henceforth refer to the service as "Approach."

An important point here: The fact that Approach has given you a discrete squawk and asked you to ident *does not* constitute clearance into the TCA. Whatever you do, don't plunge merrily along. Level off to avoid busting the 2000-foot floor, circle, do S-turns, or anything else that will keep you out of the sacred area until you hear something like this:

App: *Cherokee Six One Tango, radar contact. Cleared into the TCA. Turn right, heading zero four zero. Climb and maintain three thousand five hundred.*

You: Roger, understand cleared into the TCA. Right to zero four zero. Out of one thousand eight hundred for three thousand five hundred. Cherokee Six One Tango.

A word about altitude reporting: When working with Departure or Approach, *always* advise when you're leaving one altitude for a newly assigned altitude. The only time you don't have to worry about such reports is when you are advised to maintain VFR altitudes *at your discretion*. "At your discretion" means that you can climb or descend as you desire. In a controlled environment, however, you must report altitude changes and maintain the last one assigned. Unless specifically requested by the controller, you are not *required* to report reaching your new altitude, although some controllers prefer that you do. Use your judgment, based on frequency congestion and whether or not you have a Mode C encoder (required in TCAs and, effective December 30, 1990, in ARSAs) that automatically reports your alititude. If for any reason you can't stay VFR at the assigned altitude, advise Approach or Departure and request a new altitude. Jumping around horizontally or vertically on your own in a TCA is a very, very *verboten*. Unless you hear "at your discretion," stick to the altitude and heading assigned.

To continue the illustration:

You: Approach, Cherokee Six One Tango level at three thousand five hundred.

App: *Cherokee Six One Tango, roger. Traffic at twelve o'clock, three miles southeast bound at three thousand.*

You: Negative contact. Cherokee Six One Tango.

Perhaps a couple of minutes later:

App: *Cherokee Six One Tango, traffic no longer a factor. Climb and maintain seven thousand five hundred.*

You: Roger, out of three thousand five hundred for seven thousand five hundred. Cherokee Six One Tango.

When at the assigned altitude:

You: Approach, Cherokee Six One Tango level at seven thousand five hundred.

App: *Cherokee Six One Tango. Roger.*

When clear of the TCA, Approach will advise you accordingly:

App: *Cherokee Six One Tango, position one five miles northeast of St. Louis, departing the TCA. Squawk one two zero zero. Radar service terminated. Frequency change approved.* [If his or her workload allows, the controller might instead offer traffic advisories within the radar coverage area beyond the TCA.]

You: Roger. Cherokee Six One Tango. Thanks for your help. Good day.

DEPARTING AN ARSA

An ARSA departure offers two possible advantages for the VFR pilot over a TCA. First, if you have the traffic ahead of you (or the traffic you are supposed to follow) in sight and can maintain visual separation, Departure may clear you to turn on course and climb to your cruising altitude. In other words, once that clearance is issued, your departure is much like departing a typical, non-radar, tower-controlled airport. There would be minimum vectoring, if any, and ATC would contact you only to give you advisories of other traffic that might be in your vicinity. The principle of visual separation for VFR aircraft would be in operation. All of this means, of course, that your flight would be more direct and the departure from the ARSA more rapid.

This may not always be the standard procedure. Much will depend on weather conditions, visibility, and the volume of traffic in the ARSA. It is, however, a prerogative of ATC, when circumstances permit.

A second advantage for the VFR pilot who wants to get up and going is the size of the ARSA. The ARSA has only a 10 mile radius to the limits of the outer circle, compared to 15-30 miles for a TCA or TRSA. Consequently, you clear the regulated airspace much more rapidly and are free to fly on your own or contact Center, if you wish.

Clearance Delivery

Most ARSAs, but not all, have Clearance Delivery (the frequency is in the *Airport/Facility Directory*). If no Clearance Delivery exists, merely contact Ground Control and communicate your intentions. Ground will provide the same service as Clearance Delivery and will then clear you to taxi.

Whichever the case, the radio calls are identical. Listen to the ATIS, and then tune to the Clearance frequency. To illustrate the communication one more time, though, let's say you're leaving the Birmingham, Alabama, ARSA for Memphis:

> **You:** Birmingham Clearance, Cherokee One Four Six One Tango at the Beechcraft ramp, VFR Memphis via Victor 159. Request six thousand five hundred.

> *CD:* *Cherokee Six One Tango, Clearance. Fly runway heading and maintain three thousand. Expect six thousand five hundred in ten minutes. Departure frequency one two four point five. Squawk two six three six.*

> **You:** Roger. Runway heading, three thousand, six thousand five hundred in ten, one two four point five, and two six three six. Cherokee Six One Tango.

> *CD:* *Cherokee Six One Tango, readback correct. Contact Ground.*

Next, the usual call to Ground and then to Tower.

The Tower and Departure Control

You're at the hold line and ready to go:

You:	Birmingham Tower, Cherokee One Four Six One Tango ready for takeoff.
Twr:	*Cherokee Six One Tango, cleared for takeoff. Fly runway heading.*

In this transmission, the Tower may instruct you to "contact Departure" when airborne. If not, be sure to request the frequency change after you are off the ground and have the aircraft under control. Don't go from one frequency to another in a controlled area without permission.

The call to Departure is much the same as in a TCA or TRSA:

You:	Birmingham Departure, Cherokee One Four Six One Tango is with you, out of one thousand two hundred for three thousand.
Dep:	*Cherokee Six One Tango, radar contact. Turn right heading three one zero. Report reaching three thousand.*
You:	Right to three one zero and report three thousand. Cherokee Six One Tango.

A few minutes later:

You:	Birmingham Departure, Cherokee Six One Tango level at three thousand.
Dep:	*Cherokee Six One Tango, roger. Climb and maintain six thousand five hundred.*
You:	Roger, out of three thousand for six thousand five hundred. Cherokee Six One Tango.

When at altitude:

You:	Birmingham Departure, Cherokee Six One Tango level at six thousand five hundred.
Dep:	*Cherokee Six One Tango, roger.*

When you pass the 10-mile outer circle, you can request termination of radar advisories, if you desire. Otherwise, radar service continues until you reach the 20-mile perimeter of the outer area, when you hear the final word from Departure:

Dep:	*Cherokee Six One Tango, position twenty miles northeast of Birmingham, radar service terminated. Squawk one two zero zero. Change to advisory frequency approved.*
You:	Roger, Cherokee Six One Tango. Good day. [Be sure to change the transponder to 1200.]

In between these exchanges could have come vectors, advisories, or even safety alerts, depending on conditions or traffic within the ARSA. Or, as mentioned earlier, if the visibility is good and the traffic light, Departure Control has the prerogative to authorize a departing VFR aircraft to proceed directly on course and to continue its climb to the desired cruising altitude. That does not obviate the need for the pilot to maintain radio contact with Departure until he is clear of at least the outer circle's 10-mile radius. In the outer *area*, which is not part of the ARSA, radar service can be declined by the pilot but not by ATC.

Departing a Satellite Airport Within the Inner Circle

If you're departing a non-tower airport that's, let's say, four miles from the primary airport, and you can't reach the primary airport's tower by radio, you can still take off, but you must follow the established traffic pattern for the satellite, and, if you can't reach the tower over the radio from the ground, you must contact it as soon as practicable after takeoff. Mode C is also required, effective December 30, 1990.

DEPARTING A TRSA

Leaving a TRSA is much the same as a TCA departure, but a little simpler. As we said in Chapter 9, some (but not all) TRSAs have Clearance Delivery. However, all provide Stage III service.

Let's say we're taking off from Huntington, West Virginia, which has Clearance Delivery, and we're heading south to Knoxville. After monitoring ATIS, tune to Clearance on 118.05.

> **You:**   Huntington Clearance, Cherokee One Four Six One Tango, VFR Knoxville. Request six thousand five hundred and Stage Three.
>
> *CD:*   *Cherokee Six One Tango, Clearance. Fly runway heading, climb and maintain three thousand. Expect six thousand five hundred ten minutes after departure. Departure Control frequency one one niner point seven five. Squawk zero five two seven.*
>
> **You:**   Roger, understand runway heading, three thousand, six thousand five hundred in ten, Departure one one niner point seven five, squawk zero five two seven. Cherokee Six One Tango.
>
> *CD:*   *Cherokee Six One Tango, readback correct. Contact Ground point niner.*
>
> **You:**   Roger, Cherokee Six One Tango.

As a reminder again, enter the 0527 code in the transponder, leaving it on STANDBY. Now call Ground Control:

> **You:**   Huntington Ground, Cherokee One Four Six One Tango at the Flying Service with clearance and Information Echo.
>
> *GC:*   *Cherokee One Four Six One Tango, taxi to Runway Two One.*

The usual call to the tower at the hold line:

You: Huntington Tower, Cherokee One Four Six One Tango ready for takeoff.

Twr: *Cherokee Six One Tango, cleared for takeoff. Fly runway heading.*

Remember that the tower knows your intentions, so its instructions to you are minimal. You respond, "Roger. Cleared for takeoff. Cherokee Six One Tango," and switch your transponder to the ON or ALT position. (While Mode C is not required in TRSAs, if your aircraft is so equipped, you must have Mode C turned on.)

Once in the air, the tower will probably authorize a frequency change to Departure. If not, request the change; and call Departure:

You: Huntington Departure, Cherokee One Four Six One Tango is with you, out of one thousand five hundred for three thousand.

Dep: *Cherokee Six One Tango, radar contact. Climb and maintain six thousand five hundred.*

You: Out of one thousand five hundred for six thousand five hundred. Cherokee Six One Tango.

Dep: *Cherokee Six One tango, proceed direct Newcombe VOR.*

You: Direct Newcombe. Cherokee Six One Tango.

A few minutes later:

Dep: *Cherokee Six One Tango, you're leaving the Huntington TRSA. Stand by for traffic advisories.*

You: Cherokee Six One Tango standing by.

After another few minutes:

Dep: *Cherokee Six One Tango, position fifteen miles Southwest of Huntington, radar service terminated. Squawk one two zero zero. Change to advisory frequency approved.*

You: Roger, Cherokee Six One Tango. Good day. (Be sure to change the transponder to 1200.)

DEPARTURE FROM OTHER RADAR AIRPORTS

This time you're leaving from an airport the likes of Springfield, Missouri. Like Clarksburg in the preceding chapter, there are a few Springfields around that have no TCA, ARSA, or TRSA, but offer radar approach and departure control, nonetheless. You don't have to use this service, but why not?

To avoid repetition, let's assume that you have the ATIS, have been cleared by Ground to the hold area, and are ready to contact the tower:

> You: Springfield Tower, Cherokee One Four Six One Tango ready for takeoff, north departure.

> Twr: *Cherokee Six One Tango, cleared for takeoff. North departure approved.*

You're now off the ground:

> You: Springfield Tower, Cherokee Six One Tango requests frequency change to Departure.

> Twr: *Cherokee Six One Tango, frequency change approved.*

So you thought the tower would automatically give you the Departure frequency? *Wrong*. You should have done your homework and determined the frequency before you got in your airplane. That's just one small element of pre-flight preparation—preparations that the amateur neglects and the professional doesn't forget. But you've screwed up a little, so:

> You: Tower, Cherokee Six One Tango. What is the Departure frequency, north?

> Twr: *Cherokee Six One Tango, contact Departure on one two one point one.*

> You: One two one point one. Cherokee Six One Tango. Thank you.

> You: Springfield Departure, Cherokee One Four Six One Tango is four north of Springfield, out of two thousand for six thousand five hundred, VFR Kansas City, squawking one two zero zero. Request traffic advisories.

> Dep: *Cherokee Six One Tango, squawk zero two zero zero and ident.*

> You: Cherokee Six One Tango squawking zero two zero zero.

> Dep: *Cherokee Six One Tango, radar contact. Report reaching six thousand five hundred.*

From here on, Departure may vector you, alert you to other aircraft, or do whatever else is necessary to clear you from the area. When you reach your altitude, however, don't forget to communicate that fact:

> You: Springfield Departure, Cherokee Six One Tango. Level at six thousand five hundred.

Dep: *Cherokee Six One Tango, roger.*

Eventually, you'll hear this:

Dep: *Cherokee Six One Tango, radar service terminated. Squawk one two zero zero. Change to advisory frequency approved.*

You: Roger. Cherokee Six One Tango. Good day.

Now you're free to turn to any radio frequency that you want—Flight Service, Flight Watch, UNICOM of the airport(s) you'll be passing over a tower or ATIS of a larger airport on your route, etc. the choice is yours.

DEPARTING A NON-RADAR APPROACH/DEPARTURE CONTROL AIRPORT

Finally, you're leaving Columbia, Missouri, one of about 15 airports remaining in the U.S. with a *non-radar* Approach and Departure Control service (see Chapter 9). If you want to utilize the service, much of the accuracy of instructions relayed to you will depend on *your* accuracy in identifying your position and altitude. Keep in mind that the controller is *mentally* plotting your progress based on time, speed, and distance.

If the service is so non-precise (which it is), why use it? Well, maybe it's a hazy day with reduced, but still VFR, visibility; perhaps it's a Sunday afternoon and the surrounding atmosphere is filled with gawking sightseers at 3000 feet; or the local university has just played a major football game and some of the happy ticket-holders are homeward bound in their little singles or twins. In such situations, you should be asking for all the help you can get from the best source available. Thus the value of even non-radar control.

To get the service, make the routine tower contacts, and after takeoff, request the change to Departure frequency. Then:

You: Columbia Departure, Cherokee One Four Six One Tango, four miles west at one thousand seven hundred, climbing to six thousand five hundred, VFR Topeka. Request traffic advisories.

Dep: *Cherokee Six One Tango. Report over Boonville and level at six thousand five hundred. Traffic is a Mooney inbound over McBaine, descending from niner thousand five hundred.*

You: Roger. Report over Boonville and six thousand five hundred. Negative contact on the Mooney. Cherokee Six One Tango.

A little later, you spot what appears to be the Mooney:

> You: Departure, Cherokee Six One Tango has the Mooney at two o'clock, low.
>
> *Dep:*   *Cherokee Six One Tango. Roger.*

You're now at your cruise altitude:

> You: Departure, Cherokee Six One Tango level at six thousand five hundred.
>
> *Dep:*   *Cherokee Six One Tango, roger. Report over Boonville.*

And then:

> You: Departure, Cherokee Six One Tango over Boonville at six thousand five hundred.
>
> *Dep:*   *Cherokee Six One Tango, roger. No other traffic reported. Frequency change approved.*
>
> You: Roger. Cherokee Six One Tango. Good day.

USING CENTER OR ANOTHER AIRPORT FOR DEPARTURE CONTROL

If you browse through the *Airport/Facility Directory* you will find that many smaller airports without their own Approach/Departure Control are provided similar, but often limited, service by an Air Route Traffic Control Center or by the Approach/Departure Control of a larger airport nearby. FIGURES 9-15 and 9-16 were examples. Such service may be radar or non-radar based, depending primarily upon the distance to the radar site.

The communications required to use Center for approach/departure services are essentially the same as those discussed in the next chapter, which deals with en route VFR advisories from Center. But be aware that Center will provide this service to VFR aircraft on a workload-permitting basis only.

The procedures for using another airport's radar for your departure advisories are fundamentally the same as those covered in "Departure from Other Radar Airports" earlier in this chapter.

Remember that you can take advantages of these services, when available, even if you are operating from an uncontrolled (non-tower) field.

Thus ends our discussion of departures under the various types of control. Nothing complicated, is there? It's just a matter of knowing what to expect and what to say. If you're organized and prepared, obtaining pre-taxi clearances and using Departure is a simple process, whether you're in a TCA or leaving a non-radar airport.

Now let's turn to Approach.

APPROACH TO A TCA AIRPORT

The various radio procedures should be fairly well implanted in your mind by now, and those required in entering a TCA are only minor variations of several that we have already illustrated. Regardless, we might as well start at the beginning just to be sure that the entire range of communications has been covered.

Before calling on Approach for radar assistance into any airport, you have to be prepared to supply certain information. That information can be easily remembered if you use the acronym IPAIDS (eye-paids):

I   =   Aircraft *Identification*—type and full N number
P   =   *Position*—where you are geographically
A   =   Present *Altitude*
ID   =   *Intentions* and/or *destination*
S   =   *Squawk*—what code you are currently squawking

Keeping this sequence in mind, and rehearsing what you're going to say before getting on the mike, will convey an image of competence—which, in turn, will increase the likelihood of receiving the assistance that you want. Controllers like to deal with pilots who sound as though they know what they're doing.

Let's use St. Louis as an example again. You're coming in from the west and want to land at Lambert Field, a TCA airport. You're over the Foristell VOR, which is about 30 statute miles from Lambert and about ten miles from the outermost TCA ring. Now is the time to check the ATIS and call Approach:

> **You:**   St. Louis Approach, Cherokee One Four Six One Tango.
>
> *App:*   *Cherokee One Four Six One Tango, St. Louis Approach.*
>
> **You:**   Cherokee One Four Six One Tango, over Foristell VOR at seven thousand five hundred, landing Lambert. Squawking one two zero zero with Information Lima.
>
> *App:*   *Cherokee Six One Tango, squawk zero two three five and ident. Remain outside the TCA until radar contact.*
>
> **You:**   Cherokee Six One Tango squawking zero two three five.

Remember that just being given a squawk code does *not* authorize you to enter the TCA nor does the controller's statement that radar contact has been established. It is *only* when you hear you are "cleared into the TCA" that you can penetrate that airspace. The word "cleared" is the key. The FAA is very emphatic on this issue.

> **App:** *Cherokee Six One Tango, radar contact. Cleared into the TCA. Descend and maintain four thousand five hundred.*

> **You:** Roger, understand cleared into the TCA. Out of seven thousand five hundred for four thousand five hundred, Cherokee Six One Tango.

A few minutes later:

> **You:** Approach, Cherokee Six One Tango level at four thousand five hundred.

> **App:** *Cherokee Six One Tango. Roger. Turn left, heading zero eight five.*

> **You:** Left to zero eight five, Cherokee Six One Tango.

From this point on, Approach may give various instructions relative to new headings, altitude changes, alerts of other traffic in or near your line of flight, and the like. Each instruction or alert message should be promptly acknowledged—and not just with "roger." Remember to repeat tersely whatever the instruction is or to advise Approach if you see or don't see the traffic that is in your vicinity.

Eventually, after you've been vectored and granted permission to descend further, Approach will turn you over to the tower:

> **App:** *Cherokee Six One Tango, contact St. Louis Tower on one one eight point five.*

> **You:** Roger. One one eight point five. Cherokee Six One Tango.

Now what do you tell the tower? Almost nothing. The controller knows you're coming because Approach has so advised him. Thus there's no need for a time-consuming position/altitude/squawk report.

> **You:** St. Louis Tower, Cherokee One Four Six One Tango is with you, level at two thousand.

> **Twr:** *Cherokee Six One Tango. Enter left downwind for Runway Three Zero Left.*

> **You:** Roger, left downwind for Three Zero Left. Cherokee Six One Tango.

The instructions and landing continue in the normal pattern. To repeat some admonitions:

✈ Do not leave assigned altitudes in a TCA without permission.

✈ Always report when leaving an altitude for a newly assigned altitude. Report reaching the new altitude if you are instructed to do so (otherwise this is optional).

✈ "At pilot's discretion" means to climb or descend when you wish.

✈ When told to "ident," you don't need to acknowledge with "Cherokee Six One Tango identing." Just push the button. *Don't* push the button unless you are instructed to ident.

✈ Don't change from an assigned squawk back to 1200 until you are told.

✈ If, for any reason, you don't want to remain at the assigned altitude—or can't because of weather—request permission to climb or descend:

You: Approach, Cherokee Six One Tango requests lower.

App: *Cherokee Six One Tango, descend and maintain two thousand five hundred.*

You: Leaving three thousand five hundred for two thousand five hundred, Cherokee Six One Tango.

One other point is appropriate to emphasize in this discussion: When Approach, Departure, or Tower gives you an instruction, obey it immediately. For example, Approach tells you to turn to a heading of 120 degrees. As soon as you hear the heading, begin your turn. Don't keep flying on your present course while you pick up the mike (perhaps having to wait until the air is clear) and acknowledge the instruction. Do it *now*. Then, while in the turn, call Approach and confirm the 120-degree heading. Similarly, if you're told to climb or descend, obey immediately and confirm your action as soon as you can. The controller probably has a very good reason for varying your line of flight, so don't hesitate to do what you're told.

APPROACHING AN AIRPORT THAT UNDERLIES A TCA

Instead of landing at Lambert Field, you want to go into Spirit of St. Louis—which, you will recall, underlies Lambert's TCA to the southwest. This time, however, you're coming in from the northeast and would like vectors *through* the TCA rather than detouring around it. Despite the fact that you're landing at a non-TCA airport, the radio communications are much the same.

To set the scene, you're cruising at 6500 feet and homing on the St. Louis VORTAC, which is about 10 statute miles northwest of Lambert. Well outside the TCA, you make your first call:

You: St. Louis Approach, Cherokee One Four Six One Tango.

App: *Cherokee One Four Six One Tango, St. Louis Approach.*

You: Approach, Cherokee One Four Six One Tango is over Bunker Hill on the St. Louis zero five niner radial, level at six thousand five hundred, landing Spirit. Squawking one two zero with Information Papa.

App: *Cherokee Six One Tango, squawk zero two five two and ident. Remain outside the TCA until radar contact.*

You: Cherokee Six One Tango squawking zero two five two.

App: *Cherokee Six One Tango, radar contact three zero miles east of the St. Louis VOR. Cleared into the TCA. Descend and maintain four thousand five hundred. Turn left, heading one niner zero.*

You: Roger, understand cleared into the TCA. Out of six thousand five hundred for four thousand five hundred, left to one niner zero. Cherokee Six One Tango.

A little later:

App: *Cherokee Six One Tango, turn right, heading two seven zero.*

You: Right to two seven zero, Cherokee Six One Tango.

App: *Cherokee Six One Tango, descend and maintain two thousand five hundred.*

You: Out of four thousand five hundred for two thousand five hundred. Cherokee Six One Tango.

As you near the Spirit airport, you'll be below the 3000-foot TCA floor. About this time, Approach will call you:

App: *Cherokee Six One Tango, position five miles east of Spirit. Departing the TCA. Squawk one two zero zero. Radar service terminated. Contact Spirit Tower, one one eight point three.*

You: One one eight point three. Thank you for your help. Cherokee Six One Tango.

Now contact the Spirit tower for pattern and landing instructions: "Spirit Tower, Cherokee One Four Six One Tango is with you five miles east, level at two thousand five hundred."

Using the same situation of an airport underlying a TCA, let's suppose you are not familiar with the airport and there is no VOR or nondirectional beacon (NDB) on the field for guidance. In other words, you need assistance.

You're again coming into the St. Louis TCA from the northeast. After radar contact has been established and you are cleared into the TCA, at a given point in time, Approach calls you: "Cherokee Six One Tango, cleared *on course* [meaning "resume your own navigation]. Descend and maintain two thousand five hundred."

Being uncertain of the location of the airport, you request further help:

You: Roger. Cherokee Six One Tango is out of three thousand five hundred for two thousand five hundred. Request radar vectors to Spirit.

App: *Cherokee Six One Tango, roger. Turn right, heading two seven zero to Spirit.*

You: Right to two seven zero. Cherokee Six One Tango.

App: *Cherokee Six One Tango, Spirit is at twelve o'clock, eighteen miles. Report when you have it in sight.*

You: Will do, Cherokee Six One Tango.

In a few minutes:

You: Approach, Cherokee Six One Tango has Spirit in sight.

Approach then terminates radar service and turns you over to Spirit Tower.

Another situation isn't unusual when the TCA is busy and your destination is an underlying airport: permission to enter the TCA is not granted. In that case, getting into a field such as Spirit from the northeast means skirting the TCA. It will add time and mileage, but you have no alternative:

You: St. Louis Approach, Cherokee One Four Six One Tango.

App: *Cherokee One Four Six One Tango, St. Louis Approach.*

You: Cherokee One Four Six One Tango is over Bunker Hill, level at six thousand five hundred landing Spirit, squawking one two zero zero with Information Papa.

App: *Cherokee Six One Tango, remain clear of the TCA, squawk zero two six three and ident.*

You: Roger. Cherokee Six One Tango clear of the TCA, squawking zero two six three.

The fact that you were told to "remain clear of the TCA" is a pretty good indication that later permission is unlikely. So your only alternative is to detour around it or stay beneath the floors until the airport is in sight. At that point, call Approach once more:

You: Approach, Cherokee Six One Tango has Spirit in sight. Request change to tower.

App: *Cherokee Six One Tango, frequency change approved.*

You: Cherokee Six One Tango, thank you.

TRANSITING A TCA

This time, you don't want to land in or under the TCA but you are Mode C equipped and want to pass through it to a more distant destination. What you say and what you do are almost identical with the landing procedures. Assume that you're east of St. Louis and are heading for Jefferson City, Missouri. The straight line from your present position would put you right through the TCA. You could go around the whole thing, but, again, that would only add time and fuel costs—so why not take the economical way out? That being the more intelligent avenue, you plan what you're going to say and call Approach:

You: St. Louis Approach, Cherokee One Four Six One Tango.

App: *Cherokee One Four Six One Tango, St. Louis Approach.*

You: Approach, Cherokee One Four Six One Tango approaching Troy VOR on zero seven six radial, level at six thousand five hundred to Jeff City. Squawking one two zero zero. Request vectors through the TCA.

App: *Cherokee Six One Tango, squawk zero two six four and ident. Remain outside the TCA until radar contact.*

You: Cherokee Six One Tango squawking zero two six four.

App: *Cherokee Six One Tango, radar contact. Cleared through the TCA. Maintain six thousand five hundred. Turn right, heading two seven zero.*

A pause here to explain something: you reported your position as "approaching Troy VOR on zero seven six radial." How come the "076 radial"? Approach knows, of course, where the Troy VOR is located, so it should be easy to spot your aircraft once you've squawked the discrete code. Disregarding that, however, give the radial *from* the VOR in a position report. That makes identification of your aircraft much easier for the controller. "Approaching on the zero seven six radial" alerts him to look for you on the east side of his screen. In this case the 076 radial also happens to be the centerline of Victor 12, a VOR airway. So you could alternatively report "five miles east of Troy on Victor Twelve."

If, at any time, you are asked by Center or Approach to advise what *radial* you are flying, report it in terms of the *outbound bearing* from the station. If asked your *heading*, report the actual reading on your compass or directional gyro.

Now back to the example:

> You: Roger, understand cleared through the TCA. Maintain six thousand five hundred, right to two seven zero. Cherokee Six One Tango.

A few minutes later:

> App: *Cherokee Six One Tango, turn left, heading two four five.*
>
> You: Left to two four five. Cherokee Six One Tango.
>
> App: *Cherokee Six One Tango, traffic at eleven o'clock, two miles northeastbound at five thousand five hundred.*
>
> You: Roger, negative contact. Cherokee Six One Tango.

With your eyes scanning the skies at about the eleven o'clock position (don't forget to compensate for any crab angle), you spot the traffic approaching your position but below you:

> You: Approach, Cherokee Six One Tango has the traffic.
>
> App: *Cherokee Six One Tango, roger.*

Eventually, after whatever vectoring Approach deems necessary to provide the proper aircraft separation, you are clear of the TCA on the west side:

> App: *Cherokee Six One Tango, position two zero miles southwest of St. Louis VOR. Departing the TCA. Squawk one two zero zero. Radar service terminated. Frequency change approved.*
>
> You: Roger. One two zero zero—and thank you for your help. Cherokee Six One Tango.

You're on your own again, with Jeff City off in the distance.

APPROACH CONTROL AND THE ARSA

The mandatory Approach Control radar service at an ARSA requires no new or additional radio phraseology. The differences among TCAs, ARSAs, and TRSAs are confined to procedures or regulations:

> TCAs: *Clearance into TCA is required; clearance can be denied.*
>
> ARSAs: Radio contact with Approach *must* be established before entering the ARSA, but "clearance" into the ARSA is not required; use of radar service is mandatory.

TRSAs: No clearance or radio contact is required; Stage III service can be requested or rejected by you (''Negative Stage III'') before or after entering the TRSA.

If you keep IPAIDS in mind and have the basic TCA phraseology mastered, that's all you need when approaching an ARSA.

For example, you're headed into the Orlando International ARSA from the northwest. About 20 NM out, you're over Lake Apopka, which is identified as a reporting point on the Sectional by a small red flag:

You: Orlando Approach, Cherokee One Four Six One Tango over Lake Apopka, at three thousand five hundred, landing International. Squawking one two zero zero with Information Delta.

Now, suppose Approach comes back:

App: *Cherokee One Four Six One Tango, Stand by.*

What do you do now? The answer is: proceed on course to International. You have established the required ratio communications. Unless the controller doesn't respond at all to your call-up, or he responds with ''remain clear (or outside) of the ARSA, you can legally enter. Unlike a TCA, don't expect to hear the words ''Cleared into the ARSA.'' No *clearance* is given or required. Instead, you might hear:

App: Cherokee One Four Six One Tango, radar contact. Squawk three one zero five. Decend and maintain two thousand. Turn right, heading one eight zero.

No need to ask, ''Am I cleared into the ARSA?'' Just respond:

You: Roger, three zero one five. Down to two thousand and right to one eight zero. Cherokee Six One Tango.

There is one matter that warrants attention, however, and it could come into play with either a TRSA or an ARSA. You recall that a transponder (Mode 3/A with Mode C) is required to enter a TCA. Not so with TRSAs or ARSAs. So if your aircraft is not transponder equipped or the transponder is inoperative, how do you get any sort of radar service?

First, your aircraft will be picked up only by the ''raw,'' or ''primary,'' radar. The intensity of the target on the scope will then depend on the size of the aircraft, the extent of its metal, rain or thunderstorm activity, intervening objects (such as buildings or towers), and similar factors. If the aircraft is constructed of wood and fabric, the target will be considerably less prominent than with an all-metal aircraft.

For radar service to be provided, the non-transponder craft will have to be identified by one of two ways:

⚔ *Position Correlation:* The pilot, in his initial radio contact, communicates his position in relation to some fix, such as a VFR reporting point (denoted by a red "Visual Check Point" flag on VFR, chart), a lake, or any other feature that appears on the controller's radar map. When a correlation exists, the controller can then positively identify the aircraft.

⚔ *Identifying Turns:* Failing to establish position correlation, the controller asks the pilot to make a specific turn (30 degrees or more). The maneuver isolates and identifies that aircraft, and the pilot is then instructed to resume his course or is vectored as necessary.

To minimize radio communications, if you don't have a transponder, make that fact known in your initial call, and be as specific as possible in stating your position.

You: Orlando Approach, Cherokee One Four Six One Tango over Lake Apopka at three thousand five hundred, landing International with Information Delta. Negative transponder.

As you're just entering the ARSA's outer area, radar service will begin and continue until you're sequenced with other traffic and turned over to the Orlando Tower for landing instructions.

TRANSITING AN ARSA

If you're on a cross-country at normal cruising altitudes in the 6500 to 8500 foot range, you won't have much trouble flying above most ARSAs. Their 4000-foot AGL ceiling makes this relatively easy—unless, of course, weather conditions force a lower altitude. Out in the mountainous areas, however, it could be a different matter. Take Salt Lake City, for example. The field elevation is 4226 feet, so the ARSA ceiling would be 8200 feet AGL or higher. That elevation could pose some problems for the smaller light aircraft, especially during the summertime. Hence, if a Salt Lake or any other ARSA is in your way, contact with approach is required. Fortunately, that contact is almost identical to that when transiting a TCA. To illustrate the call, and to raise a point not previously mentioned, let's say you're en route from Memphis to the Peachtree Dekalb Airport near Atlanta and you meet Mode C requirements. The ceilings are such that you are cruising at 3500 feet MSL and are nearing the Birmingham area.

You: Birmingham Approach, Cherokee One Four Six One Tango.

App: *Cherokee One Four Six One Tango, Birmingham Approach.*

You: Approach, Cherokee One Four Six One Tango is twenty northwest on Victor One Five Niner, level at three thousand five hundred, en route Peachtree Dekalb, squawking one two zero zero. Request advisories through the ARSA.

App: *Roger, Cherokee Six One Tango. Squawk four two five three and ident.*

You: Cherokee Six One Tango squawking four two five three.

App: *Cherokee Six One Tango, radar contact.*

What happens for the next few minutes will depend on the volume of traffic in the ARSA and the advisories or safety alerts ATC issues you. But what would you do if ATC, for good reasons, asks you to leave your present altitude and climb to one that would put you in or close to the base of the overcast ceiling? You'd then be less than 500 feet below the cloud layer and in violation of VFR regulations.

The simple answer is: Don't blindly follow orders that would result in a violation. Advise ATC of the situation. The FARs make it very clear that *you're* in command of the aircraft. Yes, directions should be obeyed, but ATC may not be aware that an altitude change would make you a rule-breaker. Consequently, if any instruction would cause you to violate a regulation, regardless of ARSAs, TCAs, or otherwise, make that fact known to the controlling agency.

For altitude separation purposes between a VFR and an IFR aircraft, you might be told to climb or descend to a non-VFR cruising altitude, such as 3000 or 4000 feet. When the altitude separation is no longer needed, and especially when leaving the regulated airspace, ATC will advise you to "Resume appropriate VFR altitudes." That's your directive to climb or descend to the odd- or even-plus-500-feet VFR cruising altitudes dictated by FAR 91.109 (91.159).

Returning to the Birmingham ARSA exchanges, when you depart the Outer Area on your way to Atlanta, you'll eventually hear:

App: *Cherokee Six One Tango, position twenty miles east of Birmingham, radar service terminated. Squawk one two zero zero. Change to advisory frequencies approved.*

You: Roger, Cherokee Six One Tango, good day. (Now change the transponder to 1200.)

APPROACH CONTROL AND THE TRSAs

Keep in mind the basic difference between a TCA, ARSA, and a TRSA. You *must* have permission to enter a TCA; you must be in radio contact to enter an ARSA; but you can enter a TRSA at any time with or without permission, and use of the radar facilities is optional on your part. Otherwise, the areas are almost identical in terms of radio phraseology. To illustrate:

You: Huntington approach, Cherokee One Four Six Tango.

App: *Cherokee One Four Six One Tango, Lincoln Approach.*

You: Approach, Cherokee One Four Six One Tango is over Crown City, level at six thousand five hundred, landing Huntington squawking one two zero zero with Information Charlies.

App: *Cherokee Six One Tango, squawk zero four two five and ident.*

You: Cherokee Six One Tango squawking zero four two five.

App: Cherokee Six One Tango, radar contact. Descend at pilot's discretion to three thousand five hundred. Turn right, heading two seven zero.

You: Roger. Heading two seven zero, our discretion to three thousand five hundred. Cherokee Six One Tango.

As you're just entering the TRSA, radar service will begin and continue until you're sequenced with other traffic and turned over the Huntington Tower for landing instructions.

From this point on, it's the same old story. Just follow Approach's instructions, acknowledge them tersely, and wait until Approach turns you over to the tower frequency. That's all there is to it.

TRANSITING A TRSA

The procedures and radio communications are the same as for a TCA. However, in a TRSA you can request radar service through the TRSA after you're in it, if you decide you want that service. Also in a TRSA, you can terminate radar service at any point, and you cannot be denied entry into a TRSA. "Pilot participation is urged but is not mandatory," whether transiting, landing, or taking off. But, as we've tried to emphasize on several occasions, you're really a little foolish if you don't make use of radar surveillance and assistance in a TRSA, especially if you have a transponder.

APPROACH CONTROL: NO TCA, ARSA OR TRSA

At airports without a TCA, ARSA, or TRSA you have the option to use or not use Approach Control. The Clarksburg situation we described earlier falls in this category, as do the Designated (Mode C) Airports listed in FAR 91 Appendix D (FIG. 9-17).

Suppose you're going into Clarksburg and are unfamiliar with the area. In addition to receiving traffic advisories, you'd like to have vectors to the airport, so you call Approach:

You: Clarksburg Approach, Cherokee One Four Six One Tango.

App: *Cherokee One Four Six One Tango, Clarksburg Approach.*

You: Approach, Cherokee One Four Six One Tango is over New Martinsville level at five thousand five hundred, landing Clarksburg, squawking one two zero zero.

App: *Cherokee Six One Tango, squawk zero two zero zero and ident.*

You: Cherokee Six One Tango squawking zero two zero zero. Request.

App: *Cherokee Six One Tango, go ahead.*

You: Am unfamiliar with the area and request radar vectors to Clarksburg.

App: Cherokee Six One Tango, radar contact 35 miles northwest of Clarksburg. Fly heading of one three zero, vectors to Clarksburg.

You: Roger, one three zero. Cherokee Six One Tango.

App: *Cherokee Six One Tango, turn left, heading one two zero, descend at pilot's discretion, maintain two thousand five hundred.*

You: Left to one two zero, leaving five thousand five hundred for two thousand five hundred. Cherokee Six One Tango.

You: Approach, Cherokee Six One Tango level at two thousand five hundred.

App: *Cherokee Six One Tango, roger. Clarksburg is at twelve o'clock, seven miles. Report the airport in sight.*

You: Approach, Cherokee Six One Tango has the airport.

App: Cherokee Six One Tango, roger. Contact Clarksburg Tower one two six point seven.

You: One two six point seven. Thank you. Cherokee Six One Tango.

You: Clarksburg Tower, Cherokee One Four Six One Tango is with you, level at two thousand five hundred.

> *Twr:*   *Cherokee Six One Tango, cleared for the straight-in approach, Runway One Three. Winds one six zero degrees at seven. Altimeter two niner eight five.*
>
> You: Cleared straight in for One Niner. Cherokee Six One Tango.

NON-RADAR APPROACH

The last category is the airport that provides Approach and Departure service, but of a non-radar nature. Again, it's imprecise but better than no help at all. The only two differences between this and a radar-controlled approach are that you may be asked for more frequent position reports, and there's no point in advising your squawk. The squawk is meaningless without ground radar.

> You: Gulf Approach, Cherokee One Four Six One Tango.
>
> *App:*   *Cherokee One Four Six One Tango, Gulf Approach.*
>
> You: Approach, Cherokee One Four Six One Tango, over Pritchard, level at six thousand five hundred, landing Gulf.
>
> *App:*   *Cherokee Six One Tango, roger. Proceed direct Gulf VOR. Report ten north of the VOR.*
>
> You: Roger. Report ten north of Gulf VOR. Cherokee Six One Tango.

You've rogered the instructions, but let's say you don't have a distance measuring equipment (DME) and you're unfamiliar with the area. How will you know when you're "ten north," and how will you know when to start your descent? The best thing to do is what we all should have learned early in the game: pull out the tattered Sectional and look for a lake, a town, smokestacks, an identifying highway, or whatever will locate the 10-mile position. When you reach that point, get on the mike and so advise Approach. Remember, the controller is mentally plotting your position. As far as descending is concerned, you're not in a TCA or ARSA, so you can do so at your own discretion. As you leave one altitude and level off at another, however, advise Approach of your actions. That's the only way the controller can exert any sort of control and advice to you and others who might be in the area.

CONCLUSION

The only challenge Departure and Approach Control pose to a pilot is the challenge of the unknown. When a pilot feels insecure or uncertain about what to do or say, he'll take any action he can to skirt a TCA or an ARSA, or avoid radio contacts in a TRSA. As a result, he won't go places he'd like to go. He'll add needless miles to an otherwise straight-line cross-country. He'll land at an out-of-the-way airport and hope that some form of transportation is available to get him to his ground destination. He'll bounce around at 2500 feet to stay under a TCA when he could have a smooth flight at 4500 feet in the TCA. And so on . . . all because contacting Approach or Departure scares him. It is, he feels, beyond his expertise, beyond his scope of knowledge.

This need not be the case. Using Approach and Departure simply means knowing what you're going to say and knowing what you can expect to hear. Once you have the IPAIDS down, the rest is just a matter of listening, acknowledging, and obeying. The controller tells you what to do. Obedience (within the limits of safety and flight rules) and keeping your eyes open are your responsibility.

"But what if I don't understand the instructions? What if they use a term I've never heard of?"

That could—and does—happen. If you don't understand an instruction or a term, ask for clarification: "Say again, please." "I didn't understand." "Am not familiar with that reporting point." "Am not familiar with the area." *Don't just roger the instruction and then hope that things will work out.* They might not, and upon landing, you might find that the tower controller will want to talk with you.

No one in Approach or Departure is going to crawl through your mike cord if you sound as though you know what you're doing and then ask that an instruction be repeated or clarified. The controller *wants* to know that you *don't* know, and he wants to know when there is no uncertainty in your mind about what you are to do. That's his job—and that's one reason why instructions are repeated back. Controllers will do everything in their power to help you, guide you, and lead you safely to wherever you're going. It's up to you to give them that chance.

So use the services that you're urged to use as a VFR pilot. Know what you're going to say; say it with confidence; say it tersely but distinctly; say it with the ring of a professional. As so many pilots have learned, you'll get the help you need.

11

Centers

M ost of the previous chapters have kept us in the general vicinity of the air-port—Tower, Approach, Departure, and the like. Now is the time to stray beyond the local aerodrome or TCA limits and consider the pilot's good cross-country friend: The Air Route Traffic Control Center (ARTCC, or "Center," for short).

CENTER'S ROLE IN THE SCHEME OF THINGS

Its name identifies its primary purpose and role. Center is indeed the *center* for control of the flow of air traffic over the routes within its assigned geographical area. It ensures the vertical and horizontal separation of all IFR aircraft and alerts participating VFR aircraft to potential traffic hazards. In performing its task, *control* is the critical word. That is Center's job—which means that functions not directly related to the control of traffic, such as filing or amending a flight plan, requesting en route weather, and the like, should be funneled through Flight Service.

In a realistic sense, Center is the middleman between Departure Control and Approach Control. Once clear of the TCA, ARSA, or TRSA, Departure drops out of the picture, and Center steps in. For the duration of the flight, one or more Centers (depending on the distance traveled) become the en route controllers, assigning altitudes and vectors to IFR aircraft, advising VFR pilots of potential traffic, and generally monitoring the safety of the airways. As the flight nears its end, Center either terminates the radar service or hands the pilot off to Approach Control for sequencing and separation prior to the actual tower-controlled landing.

So, once again, the basic flow from taxi-out to landing is:

Ground Control
↓
Tower
↓
Departure Control
↓
Center
↓
Approach Control
↓
Tower
↓
Ground Control

THE CENTERS AROUND THE COUNTRY

Tied together by sophisticated communications networks, 23 Centers control the en route traffic in the United States. By name and general location, they are:

Albuquerque	Honolulu	Minneapolis
Anchorage	Houston	New York
Atlanta	Indianapolis	Oakland
Boston	Jacksonville	Salt Lake City
Chicago	Kansas City	San Juan
Cleveland	Los Angeles	Washington
Denver	Memphis	
Fort Worth	Miami	

Fair enough, but what about the geography between some pairs of these locations? Take Washington and Jacksonville, for example. There are 647 statute miles separating the two. How is there any control when you're several hundred miles from one center to the other? The obvious answer is remoted communications that tie the more distant sites to the physical location of the Center by land lines or microwave links.

A pilot flying from Washington to Jacksonville would first be tuned to one Washington Center "sector" frequency until advised by Center to change to another sector frequency as the flight progresses—each frequency remoted to the Leesburg (Virginia) ARTCC back near the nation's capital. As the pilot leaves Washington Center's area (south of Wilmington, North Carolina) he is so advised and given the first remoted Jacksonville Center frequency to contact. In this case, it would be Jacksonville's site at Myrtle Beach. In time, Jacksonville Center would advise the pilot to change again to a remoted site at Savannah, and so on—sector by sector—until the aircraft reaches the immediate Jacksonville vicinity.

Throughout the country, with the network of communication facilities, you can always be in touch with one Center or another (FIG. 11-1).

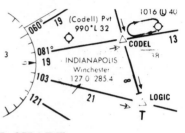

Fig. 11-1. *You may be near Winchester, Kentucky, but as the Enroute Low Altitude Chart shows, the Indianapolis Center controls this sector through its remoted Winchester outlet. The remoted VHF and UHF sector frequencies are shown.*

THE ENROUTE LOW ALTITUDE CHART

Perhaps you noticed something a little different about FIG. 11-1. Although it's only a fragment of a chart, it doesn't look at all like the familiar Sectional. That's because it's not. It's a piece of an Enroute Low Altitude Chart, FIG. 11-2.

Some mention of this chart could have come at any time in the book. However, we've reserved reference to it until now because of our discussion of Center, cross-country excursions, and ways to make such flights safer and easier. If you're not familiar with the chart, stay with us for a brief summary. If you know all about it, skip the whole section.

Twenty eight charts cover the country, each valid for a period of about two months. These are not 28 individual charts but rather two in one, as FIG. 11-3

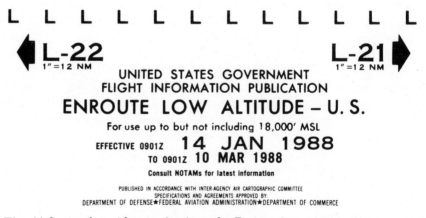

Fig. 11-2. *Another aid to navigation—the Enroute Low Altitude Chart.*

indicates. L 21 extends from western Missouri east to portions of Kentucky. L 22 overlaps L 21 in Kentucky and continues east over parts of West Virginia and Virginia to the Atlantic coast.

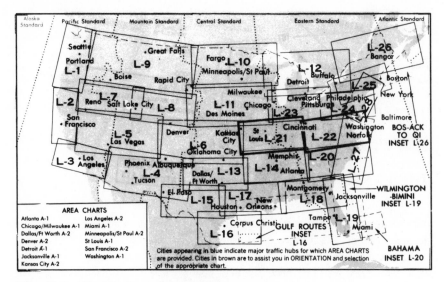

Fig. 11-3. *The 28 charts that cover the country.*

It's called "Enroute Low Altitude" because it is for use up to but not including 18,000 feet MSL. For the VFR pilot who wants to track VORs and use his radio effectively, it provides a wealth of information not found on the Sectional. FIGURE 11-4 illustrates some of the typical data. Keep in mind that it is an *en route* chart and, thus, omits some information found on the Sectional, such as UNICOM frequencies. Nor does it depict topography, towns, roads, rivers, and similar landmarks.

Despite the absence of that information, the chart can almost totally replace the Sectional during the en route portions of a flight *if* you want to rely primarily on your navcom equipment for navigation. Indeed, it is for IFR flight and contains a lot of symbols that aren't pertinent to the VFR pilot. That fact doesn't reduce its usefulness, however.

We're not suggesting that you discard the Sectional, only that you use the two jointly. After all, you might lose radio contact and have to rely on the Sectional to get you to the nearest airport. It's a good idea to use the charts in combination so that your position relative to identifiable ground references is never in question.

What do you find in the Enroute that's not included in the Sectional? A few examples are illustrated in FIGS. 11-5 through 11-9. These represent some of the more meaningful data for the VFR pilot that make the chart a helpful reference

158

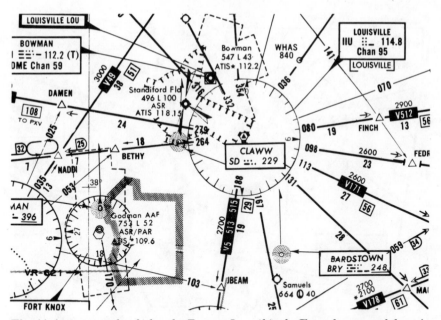

Fig. 11-4. *An example of what the Enroute Low Altitude Chart shows—and doesn't show.*

and navigation resource. It's not an essential tool, and if you're going to travel strictly by dead reckoning, leave it home. Otherwise, plan your flight with both charts, and use both as you venture from here to there.

THE ADVANTAGES OF CENTER FOR THE VFR PILOT

First, it should be made clear that en route assistance from Center is neither mandatory nor always available for the VFR pilot. The primary purpose of the ARTCC is to facilitate the movement of IFR aircraft. It does not exist nor is it required to serve the VFR pilot. You may request en route VFR advisories, but the controller has the right to refuse the request if his workload does not permit. It is very likely that in marginal VFR conditions or when there is a heavy concentration of traffic, your request for advisories will be rejected. At other times, you will often find Center's controllers not only helpful but anxious to be of help.

A second observation: A Center controller is a professional who serves professionals—meaning airline, corporate, and skilled instrument pilots. The vast majority of these airmen know (or *should* know) how to use their radios. The air-to-ground communications are brief and to the point.

When a VFR pilot gets on the air and proceeds to stammer, hesitate, ramble, or give the impression of incompetence, the likely response from Center to the request for en route advisories will be "unable due to workload." That may,

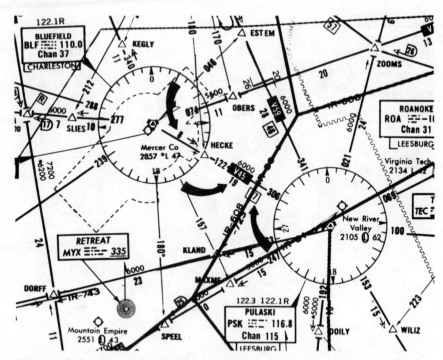

Fig. 11-5. *Nautical mileage between VORs, airway intersections, and other reporting points. In this case, eight miles between Bluefield VOR and Hecke Intersection, and 19 miles between Hecke and Pulsaki VOR. The boxed "27" (partially obscured in this reproduction) indicates the total mileage between the two VORs—just like the mileage breakdowns on a road map. You won't find mileage on the Sectionals.*

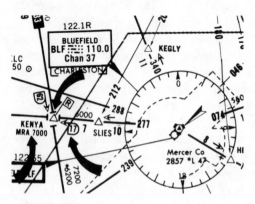

Fig. 11-6. *DME (Distance Measuring Equipment) fix (Kenya), mileage from the VOR (17 NM) to Kenya, and Minimum Reception Altitude at the fix (7000' MSL).*

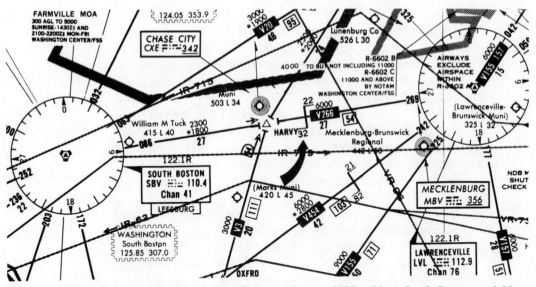

Fig. 11-7. *VOR changeover points, with mileage to the two VORs (32 to South Boston and 22 to Lawrenceville).*

Fig. 11-8. *Identification of a VOR Airway (Victor 132), Minimum Enroute Altitude for radio reception between two stations (3000 feet MSL), and Minimum Obstruction Clearance Altitude (2400 feet MSL).*

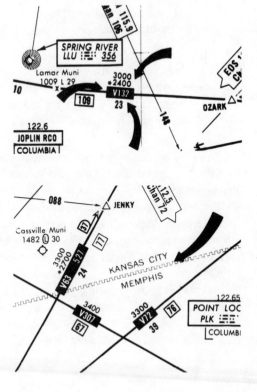

Fig. 11-9. *Air Route Traffic Control Center boundaries.*

in fact, not be the case at all, but the controller has concluded that he just doesn't have the time to fiddle around with an unprepared amateur. Nor can he be blamed. Remember the case of an airline pilot who reported his position in only five seconds while a private pilot took four minutes to convey the same basic information. Center has neither the time nor patience to accommodate that sort of amateurism.

None of this is intended to imply that the VFR pilot should avoid using Center or to intimidate the relatively inexperienced. The 100-hour pilot can sound just as professional as the 10,000-hour 747 captain. All it requires is practiced competence in the art of radio communications.

The advantages of using Center on a VFR cross-country are enough to justify the development of professionalism. Let's cite a few:

- ✈ The controller with whom you are in contact is, in effect, another pair of eyes—groundbound though they may be. He has *your* airplane on his scope, identified by the discrete transponder code he has assigned you, thus he can alert you to traffic in your general vicinity. He can issue you warnings of military flight activity, such as B-52s on low-level training flights. If you want to change altitudes, he can advise you of potential traffic at the new altitude you have chosen.

- ✈ If you lose all of your radio and squawk the 7700/7600 combination, your Center controller will spot the radio failure. Subsequent controllers will be aware of the situation.

- ✈ If you encounter a serious emergency, you have someone on the frequency to whom you can talk *now*—no retuning to 121.5; no need to divert your attention from handling the emergency to recode the transponder to 7700. The controller knows where you are and is in a position to alert the sources closest to your position of your predicament.

In every respect, Center is an added insurance policy to your flight plan. As with all of the other radio aids, the service is there to be used. We must stress again, however, that Center is not *required* to lend assistance to a VFR pilot. It's the one case where requested advisories can be rejected. The tower has no such freedom; Approach and Departure have no such freedom when you are landing or departing a TCA, ARSA, or TRSA airport; Ground Control has no such freedom. Only Center does. It must assist the IFR pilot but not those flying VFR.

As a VFR pilot, you can legally fly the length and breadth of these United States and never once contact any Center. But the proverbial question: Why not use the service, since it's available? At the very least, monitor the remoted frequencies as you pass from one area to another. Just eavesdropping may alert you to a potential hazard somewhere in your line of flight. It's just as easy, however, to go whole-hog—to go first class—if you know what you're doing.

Now we'll illustrate the typical radio contacts.

GOING FROM DEPARTURE TO CENTER

Either of two situations can occur when Departure (or Approach) Control has been vectoring you out of the TCA, ARSA, or TRSA. You can request a "handoff" to Center (meaning that the agency with which you are in contact turns you over to the next-in-line agency) and *get* the handoff. Or, your request can be denied, for whatever reason. How would the two dialogues go?

You:	Departure, Cherokee Six One Tango request.
Dep:	*Cherokee Six One Tango, go ahead.*
You:	Departure, Cherokee Six One Tango requests a handoff to Center.
Dep:	*Cherokee Six One Tango, stand by. We'll check . . .* [pause] *. . .*
Dep:	*Cherokee Six One Tango, contact Kansas City Center, frequency one two five point five five.*
You:	Roger. One two five point five five. Cherokee Six One Tango. Good day.

Inasmuch as Departure said nothing about not being able to complete the handoff, you can assume that your request has been honored. So you merely reset your radio to 125.55 and call Center:

You:	Kansas City Center, Cherokee One Four Six One Tango is with you, level at seven thousand five hundred.
Ctr:	*Cherokee One Four Six One Tango, roger. Altimeter Two Niner Two Four.*

Four points here:

✈ The expression "with you" is used any time you are being handed off or automatically transferred from one agency to another: Departure to Center; Center to Approach; Approach to Tower.

✈ Even though it is a handoff, be sure to advise the receiving agency of your present altitude. If you fail to include it in your initial contact, you may be asked—which only adds to air clutter.

✈ Note that no position report is necessary. Center, or whatever the controlling agency is, knows your position as well as your destination and squawk.

✖ Do not change your transponder from one squawk to another until you are so advised. Let's say that Departure has given you 1201, and you have been handed off to Center. Keep 1201 in the box unless Center gives you a new code.

The second situation:

> **You:** Departure, Cherokee Six One Tango request.
>
> *Dep:* *Cherokee Six One Tango, go ahead.*
>
> **You:** Departure, Cherokee Six One Tango requests a handoff to Center.
>
> *Dep:* *Cherokee Six One Tango, stand by, we'll check.* [Pause]
>
> *Dep:* *Cherokee Six One Tango, unable at this time. Radar service terminated. Squawk one two zero zero. Frequency change approved.*
>
> **You:** Roger. Squawking one two zero zero. Cherokee Six One Tango. Good day.

Any time that you hear the word *terminated,* you can assume that all radar service has ceased and that no handoff to the next agency has taken place. Now if you want Center on Approach, you have to give the full IPAIDS. The next agency hasn't heard about you before.

We should observe at this point that some controllers misuse "terminated" as it is supposed to be applied per the book. You hear something like this from Departure: "Cherokee Six One Tango, radar service terminated. Contact Kansas City Center on one two five point five five." Has radar service *literally* been terminated, or have you been handed off to center? The first part says terminated; the second part raises a question. Uncertain of what has really happened, you play the pro and call Center:

> **You:** Kansas City Center, Cherokee One Four Six One Tango.
>
> *Ctr:* *Cherokee One Four Six One Tango, Kansas City Center. Go ahead.*

The phraseology indicates that this is the initial contact with the controller, so you go ahead with IPAIDS.

But suppose you hear this in response to that call:

> *Ctr:* *Cherokee One Four Six One Tango, Kansas City Center. Squawk one two zero five and ident. Verify altitude.*

You: Cherokee One Four Six One Tango, squawking one two zero five. Level at seven thousand five hundred.

This time the phraseology tells you that you have been handed off. In either case, once contact with Center has been established, the controller will take care of much of the conversation from that point on. Your basic job is to listen, respond to, and acknowledge whatever information or instructions are conveyed.

INITIATING THE CONTACT WITH CENTER

Either Departure hasn't handed you off or you haven't used Departure to leave the area. Regardless of the situation, you want to establish the first contact with a Center. Again, it's the simple IPAIDS:

You: Kansas City Center, Cherokee One Four Six One Tango.

Ctr: *Cherokee One Four Six One Tango, Kansas City Center. Go ahead.*

You: Center, Cherokee One Four Six One Tango ten miles south of Kansas City, level at seven thousand five hundred, VFR Memphis, squawking one two zero zero. Request VFR advisories, workload permitting.

Ctr: *Cherokee Six One Tango, squawk one two zero six and ident.*

You: Cherokee Six One Tango, squawking one two zero six.

Ctr: *Cherokee Six One Tango, altimeter two niner two five.*

That's all there is to it. Center will take it from there and keep you advised of what's going on around you.

About that comment, "workload permitting." It's obviously not required, any more than "good day" or "thank you," but it does put your request in the form of a *request*—not a command. In its way, it indicates appreciation of the fact that the controller might be busy and not able to provide the advisories. That little added phrase is frequently enough to get the help that otherwise might have been rejected. Call it courtesy and mutual understanding in the air.

Now that Center has you on radar and is tracing your progress, an admonition is in order. Don't change altitudes without advising Center. The controller has you pegged at seven-point-five, and if you don't have a Mode C transponder, there's no way he can determine your altitude without verification from you. A change to a different flight level might affect the flow of traffic and place you and others in jeopardy. You're not flying under instrument flight rules, and you do have the VFR freedom to vary your altitude. But you've told Center one thing, so don't make changes without communicating your intended actions.

EN ROUTE FREQUENCY CHANGES

You're moving along over the countryside tuned to Kansas City Center when you hear something like this:

Ctr: *Cherokee Six One Tango, contact Kansas City Center, frequency one two five point three.*

You: Roger. One two five point three. Cherokee Six One Tango. Good day.

The call simply means that you're passing out of the controller's sector. So you "good day" him and tune in 125.3. Then make contact on the new frequency:

You: Kansas City Center, Cherokee One Four Six One Tango with you, level at seven thousand five hundred.

All you're doing is talking to a different controller at Kansas City Center who is handling the geographical area you are now in.

Along the same line, you might hear this:

Ctr: *Cherokee Six One Tango, change to my frequency* [or "contact me now on"] *one one eight point five five.*

You: Roger. One one eight point five five. Cherokee Six One Tango.

Change your radio and make contact again: "Kansas City Center, Cherokee Six One Tango is with you on one one eight point five five."

Notice the difference between the calls? In the latter case, the controller said ". . . change to *my* frequency . . . " It's the same person you've been chatting with all along. He just wants you on a different remote site frequency. There's no need for the traditional "good day" or altitude report. You haven't left him. Listen for the inclusion of *me* and/or *my* in the controller's message. That's the tipoff.

EN ROUTE ADVISORIES AND CENTER

Back to the flight itself: You're in tune with Center and squawking the code given you. What the controller passes on to you now is similar to the messages Approach or Departure might relay—traffic alerts and the like. There may be frequent advisories; there may be none. It all depends on the controller's workload and the conditions on your route of flight. Typical of what you might hear are these examples:

Example 1

Ctr: *Cherokee Six One Tango, traffic at ten o'clock, three miles, northeastbound. Altitude unknown.*

You: Negative contact. Cherokee Six One Tango. [Or, Cherokee Six One Tango has the traffic. Or, Cherokee Six One Tango requests vectors around the traffic.]

Some old-time aviators still use the phrase "no joy" instead of "negative contact," and "tallyho" instead of "has the traffic." However, the FAA frowns upon such phraseology because the current generation of controllers is unfamiliar with it.

Example 2

Ctr: *Cherokee Six One Tango, were you advised of low-level military flights in your present area?*

You: Negative, not so advised.

Ctr: *Cherokee Six One Tango, be alert for north-south B-52 activity.*

You: Cherokee Six One Tango will be looking. Thank you.

Example 3

You see ahead of you a cloud buildup that appears to be right at your altitude. To avoid the clouds you decide to drop down from 7500 to 5500 feet. That's permissible, but advise Center ahead of time of your plans:

You: Center, Cherokee Six One Tango descending to five thousand five hundred due to clouds.

Ctr: *Cherokee Six One Tango, roger. Report reaching five thousand five hundred.*

You: Cherokee Six One Tango wilco.

You: Center, Cherokee Six One Tango level at five thousand five hundred.

Ctr: *Cherokee Six One Tango, roger.*

Example 4

You are leaving Kansas City Center's jurisdiction and approaching that of the Memphis Center:

Ctr: *Cherokee Six One Tango, contact Memphis Center now, frequency one two four point three five.*

You: Roger. One two four point three five. Cherokee Six One Tango.
            Good day.

Note that Kansas City has said nothing about radar service being terminated.
This means that you have been handed off to Memphis. Just change to 124.35
and make your call:

You: Memphis Center, Cherokee One Four Six One Tango is with you,
            level at five thousand five hundred.

Ctr: *Roger. Cherokee Six One Tango. Altimeter three zero one five.*
            Continue present heading [or whatever the instructions, if any, might
            be].

A comment or two about frequency changing is appropriate here. Let's say
that you have two navcoms. The last Kansas City frequency, 125.3, is in navcom
#1. When told to change to Memphis, enter 124.35 in navcom #2, but leave #1
where it is. If for any reason you have to call Kansas City again, you can do
so without time-consuming dial changes. Once contact is made with Memphis,
merely set navcom #1 to the next probable Memphis remote frequency, or turn
it to any other radio aid you want.

    With only one navcom aboard, be sure to write down the new frequency
given you. (It's easy to forget or confuse 124.35 with 123.45—or any other
combination.) If you've done this progessively from the first Departure frequency
through the one or more Center frequencies, the piece of paper on your knee
pad will resemble this:

KC D/C	~~119.0~~
KC Ctr	~~125.55~~
KC Ctr	~~125.3~~
MEM Ctr	124.35
	Etc.

Just cross out each previous frequency, but don't obliterate it. You *might*
need it again.

    Even with two navcoms, this practice of writing down each succeeding
frequency makes good sense. It's a little embarrassing, after a minute or two of
uncertainty or forgetfulness, to have to recontact the last controller and ask: "What
frequency did you tell me to change to?"

Example 5

    As the flight progresses, you find that the headwinds are much stronger than
forecast. Your forward progress is slower than the rate at which the fuel gauges
are dropping, so you decide to put down at Springfield, Mo. Because you're still

in Kansas City Center's area, and they have you on a nonstop flight to Memphis, a call to Center is essential:

You: Center, Cherokee Six One Tango will be landing Springfield for fuel.

Ctr: *Cherokee Six One Tango, roger. Radar service terminated. Squawk one two zero zero. Contact Springfield Approach on one two four point niner five.*

You: Roger. One two four point niner five. Cherokee Six One Tango. Good day.

As you know from the last chapter, Springfield has no TCA, ARSA, or TRSA, so the choice is yours to contact Approach or not. When on the ground, be sure to amend your flight plan with Flight Service. Center won't do it for you.

CENTER TO APPROACH

Airborne again (or maybe you didn't have to stop at all), you're nearing Memphis International. At a given point, Memphis Center will contact you with one of three possible instructions.

Example 1

Ctr: *Cherokee Six One Tango, radar service terminated. Contact Memphis Approach on one two six point zero five.*

You: Roger. One two six point zero five. Cherokee Six One Tango. Good day.

Center said "terminated," so you can assume that you are *not* being handed off. But the controller gave no instructions about changing the transponder to 1200 or any other code, which means that you leave it at the last setting—1205, or whatever it was. Then call Approach with the standard IPAIDS:

You: Memphis Approach, Cherokee One Four Six One Tango.

App: *Cherokee One Four Six One Tango, Memphis Approach.*

You: Cherokee One Four Six One Tango is with you over Marion, level at five thousand five hundred, landing Memphis. Squawking one two zero five with Information Quebec.

Approach will take it from there.

Example 2

Ctr: *Cherokee Six One Tango, radar service terminated. Squawk one two zero zero. Contact Memphis Approach on one two six point zero five.*

You: Roger. One two six point zero five. Cherokee Six One Tango. Good day.

With this advice, you know there has been no handoff, so change to 1200, and give Approach the full IPAIDS.

Example 3

Ctr: *Cherokee Six One Tango, contact Memphis Approach on one two six point zero five.*

You: Roger. One two six point zero five. Cherokee Six One Tango. Good day.

You: Memphis Approach, Cherokee One Four Six One Tango is with you, level at five thousand five hundred.

This was a direct handoff. Approach knew all about you and your intentions— thus no need for IPAIDS, other than to confirm your present altitude.

USING CENTER AS APPROACH/DEPARTURE CONTROL

As mentioned in Chapter 10, Center offers limited Approach and Departure Control for many airports that would otherwise have no approach/departure services, including:

- ✈ non-tower airports that are not within radar coverage of a larger airport's Approach/Departure Control.

- ✈ tower airports where activity levels do not justify airport-based radar and which are not within the radar coverage of a larger airport's Approach/Departure Control.

- ✈ airports with a part-time tower and part-time Approach/Departure Control when those services are closed.

- ✈ tower and non-tower airports within radar coverage of a larger airport's part-time Approach/Departure Control when such service is closed.

Like the other Center services discussed, VFR approach and departure services are available from Center on a workload-permitting basis only. FIGURES 11-10 through 11-13 are just three examples of airports with Center-supplied approach and departure services.

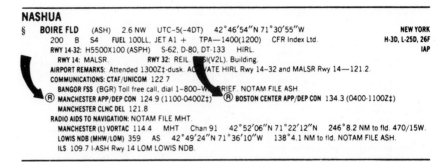

NASHUA

§ **BOIRE FLD**   (ASH)   2.6 NW   UTC-5(-4DT)   42°46'54"N 71°30'55"W   **NEW YORK**
   200   B   S4   FUEL 100LL, JET A1 +   TPA—1400(1200)   CFR Index Ltd.   **H-3D, L-25D, 26F**
   RWY 14-32: H5500X100 (ASPH)   S-62, D-80, DT-133   HIRL.   **IAP**
   RWY 14: MALSR.   RWY 32: REIL. [?]SI(V2L). Building.
   AIRPORT REMARKS: Attended 1300Z‡-dusk. AC[?]VATE HIRL Rwy 14–32 and MALSR Rwy 14—121.2.
   COMMUNICATIONS: CTAF/UNICOM 122 7
   BANGOR FSS (BGR) Toll free call, dial 1–800–W[?]RIEF. NOTAM FILE ASH.
Ⓡ MANCHESTER APP/DEP CON 124 9 (1100-0400Z‡)   Ⓡ BOSTON CENTER APP/DEP CON 134.3 (0400-1100Z‡)
   MANCHESTER CLNC DEL 121.8
   RADIO AIDS TO NAVIGATION: NOTAM FILE MHT.
   MANCHESTER (L) VORTAC 114.4   MHT   Chan 91   42°52'06"N 71°22'12"N   246°8.2 NM to fld. 470/15W.
   LOWIS NDB (MHW/LOM) 359   AS   42°49'24"N 71°36'10"W   138°4.1 NM to fld. NOTAM FILE ASH.
   ILS 109.7 I-ASH Rwy 14 LOM LOWIS NDB.

Fig. 11-10. *Pilots at Boire Field, a non-tower airport at Nashua, New Hampshire, can use the Manchester airport's Approach and Departure Control during the daytime and Boston Center at night (when the Manchester Tower is closed). The Ⓡ indicates that both ATC facilities rely on radar for this service.*

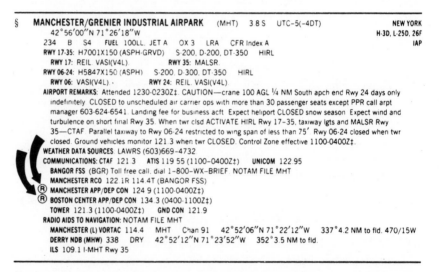

§ **MANCHESTER/GRENIER INDUSTRIAL AIRPARK**   (MHT)   3.8 S   UTC-5(-4DT)   **NEW YORK**
   42°56'00"N 71°26'18"W   **H-3D, L-25D, 26F**
   234   B   S4   FUEL 100LL, JET A   OX 3   LRA   CFR Index A   **IAP**
   RWY 17-35: H7001X150 (ASPH-GRVD)   S-200, D-200, DT-350   HIRL
   RWY 17: REIL VASI(V4L).   RWY 35: MALSR.
   RWY 06-24: H5847X150 (ASPH)   S-200. D-300, DT-350   HIRL
   RWY 06: VASI(V4L). ·   RWY 24: REIL. VASI(V4L).
   AIRPORT REMARKS: Attended 1230-0230Z‡. CAUTION—crane 100 AGL ¼ NM South apch end Rwy 24 days only
      indefinitely CLOSED to unscheduled air carrier ops with more than 30 passenger seats except PPR call arpt
      manager 603-624-6541. Landing fee for business acft Expect heliport CLOSED snow season Expect wind and
      turbulence on short final Rwy 35. When twr clsd ACTIVATE HIRL Rwy 17-35, taxiway lgts and MALSR Rwy
      35—CTAF Parallel taxiway to Rwy 06-24 restricted to wing span of less than 75' Rwy 06-24 closed when twr
      closed. Ground vehicles monitor 121 3 when twr CLOSED. Control Zone effective 1100-0400Z‡.
   WEATHER DATA SOURCES: LAWRS (603)669-4732
   COMMUNICATIONS: CTAF 121 3   ATIS 119 55 (1100-0400Z‡)   UNICOM 122.95
   BANGOR FSS (BGR) Toll free call, dial 1–800–WX–BRIEF. NOTAM FILE MHT
   MANCHESTER RCO 122 1R 114 4T (BANGOR FSS)
Ⓡ MANCHESTER APP/DEP CON 124 9 (1100-0400Z‡)
Ⓡ BOSTON CENTER APP/DEP CON 134 3 (0400-1100Z‡)
   TOWER 121.3 (1100-0400Z‡)   GND CON 121.9
   RADIO AIDS TO NAVIGATION: NOTAM FILE MHT.
   MANCHESTER (L) VORTAC 114.4   MHT   Chan 91   42°52'06"N 71°22'12"W   337°4.2 NM to fld. 470/15W
   DERRY NDB (MHW) 338   DRY   42°52'12"N 71°23'52"W   352°3.5 NM to fld.
   ILS 109.1 I-MHT Rwy 35

Fig. 11-11. *Pilots at Manchester, one of many airports with a part-time tower, have their own Approach and Departure Control during the daytime, but they too use Boston Center at night when the tower is closed.*

CONCLUSION

That's pretty much the story of Center, as far as the VFR pilot is concerned. To summarize some of the points we've made:

✈   Center exists primarily to serve IFR flights.

✈   Its assistance to VFR pilots is on a "workload permitting" basis.

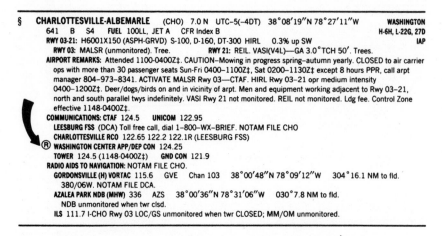

§ **CHARLOTTESVILLE-ALBEMARLE**   (CHO) 7.0 N   UTC−5(−4DT)   38°08′19″N 78°27′11″W   **WASHINGTON**
641   B   S4   **FUEL** 100LL, JET A   CFR Index B   **H-6H, L-22G, 27D**
RWY 03-21: H6001X150 (ASPH-GRVD) S-100, D-160, DT-300 HIRL   0.3% up SW   **IAP**
    RWY 03: MALSR (unmonitored). Tree.   **RWY 21:** REIL. VASI(V4L)—GA 3.0°TCH 50′. Trees.
    AIRPORT REMARKS: Attended 1100-0400Z‡. CAUTION—Mowing in progress spring–autumn yearly. CLOSED to air carrier
    ops with more than 30 passenger seats Sun-Fri 0400–1100Z‡, Sat 0200–1130Z‡ except 8 hours PPR, call arpt
    manager 804–973–8341. ACTIVATE MALSR Rwy 03—CTAF. HIRL Rwy 03–21 opr medium intensity
    0400–1200Z‡. Deer/dogs/birds on and in vicinity of arpt. Men and equipment working adjacent to Rwy 03–21,
    north and south parallel twys indefinitely. VASI Rwy 21 not monitored. REIL not monitored. Ldg fee. Control Zone
    effective 1148-0400Z‡.
    COMMUNICATIONS: CTAF 124.5   **UNICOM** 122.95
    LEESBURG FSS (DCA) Toll free call, dial 1–800–WX-BRIEF. NOTAM FILE CHO
    CHARLOTTESVILLE RCO 122.65 122.2 122.1R (LEESBURG FSS)
    ⓡ **WASHINGTON CENTER APP/DEP CON** 124.25
    TOWER 124.5 (1148-0400Z‡)   **GND CON** 121.9
    RADIO AIDS TO NAVIGATION: NOTAM FILE CHO.
    GORDONSVILLE (H) VORTAC 115.6   GVE   Chan 103   38°00′48″N 78°09′12″W   304°16.1 NM to fld.
     380/06W. NOTAM FILE DCA.
    AZALEA PARK NDB (MHW) 336   AZS   38°00′36″N 78°31′06″W   030°7.8 NM to fld.
     NDB unmonitored when twr clsd.
    ILS 111.7 I-CHO Rwy 03 LOC/GS unmonitored when twr CLOSED; MM/OM unmonitored.

Fig. 11-12. *Charlottesville, Virginia, uses Washington Center for Approach/ Departure Control when the tower is closed—*and *when it's open.*

🛩 Use of Center by the VFR pilot is *advisable* but not mandatory.

🛩 Although the VFR pilot is not under Center's control, he should not deviate from announced altitudes or routes of flight without advising Center in advance.

🛩 Become familiar with the probable frequencies you will use, and know approximately when you will be asked to change from one remoted frequency to another or from one Center to another.

🛩 Become familiar with the Enroute Low Altitude chart for more exact radio navigation and frequency-change areas.

🛩 Write down all frequencies (planned and/or given) in chronological sequence so you won't forget or become confused.

🛩 Plan your communication to Center before picking up your mike.

🛩 Rehearse your message so that you sound like a pro—not a hesitant amateur.

Center is one more eye in the sky to help you get where you want to go. We urge that you use this control source, when possible, for that added insurance. We urge that you use it on every cross-country flight. Too many non-IFR pilots, because of lack of knowledge and confidence, are afraid of it—when there is no reason for fear of any sort. If the controller can't accommodate you, he'll tell you; if you don't understand a direction, ask him to repeat or clarify. We've heard airline pilots, who are presumably pros, do this on any number of occasions.

In every case, the controller honored the request without rancor or intimidation. He'll do the same for you *if* you sound as though you know what you're doing and what's going on.

One final observation: there's an old saying that "what we're not up on, we're down on." What we don't understand, we're either against or we fear. Uncertainty breeds insecurity. This chapter (and, in fact, the whole book) tries to provide some knowledge that will lead to understanding and greater pilot confidence in the field of radio communications.

Rarely can examples and the written word accomplish the entire task, however. Consequently, we recommend that you visit the Center nearest you. Call a supervisor, explain what you want, and request a brief tour of the facility. See what's going on. Talk to a couple of controllers. Listen to the communications between ground and air. Study the screens and the blips that identify the various aircraft.

Firsthand experience is the best way we know to bring meaning to words and examples to life. The FAA urges all pilots to visit its facilities—Center, Tower, Flight Service, and the rest. The facilities are glad to welcome you, and, workload permitting, will see that your tour is complete and educational. It's worth a couple of hours of your time to put to rest any feelings of uncertainty. "What we're up on, we're not down on." The saying works in reverse as well.

12

In the Event
of Radio Failure

Despite the sophistication of modern avionics, things can go wrong. When your radio(s) decides to take a rest, it is comforting to know what you have to do to get your airborne vehicle back safely on the ground.

As the *AIM* says, it's virtually impossible to establish fixed procedures for all situations involving two-way radio failure. Preferred actions, however, can be outlined. Within those parameters, every pilot should have a plan in the unlikely event communications are lost. Just as the intelligent pilot is mentally prepared for an engine failure, so is he ready to cope with a radio failure.

ONE PREVENTIVE STEP

It's pretty hard to determine the potential for radio failure during the pre-flight check. Of course, a loose or broken antenna is an obvious signal, as is a navcom that slips in and out of its housing rack, a circuit breaker that has popped, or a frayed magneto wire. Otherwise, there's not a lot you can check.

One symptom of a potential problem, however, is the alternator belt. If you can inspect it visually, check it for wear and frays. Whether you can see it or not, check its tension. If it's loose, have it replaced or tightened. Not all radio failures can be traced to alternators or alternator belts, but malfunctioning of one or the other will result in a general power loss, with the battery supplying what little remaining electrical juice it can generate. And the battery won't last forever under those conditions.

WHEN A FAILURE IS SUSPECTED

Every once in a while, the radio can give off an eerie silence. What has been a constant pattern of chatter between the controller and other aircraft suddenly is no more. You can't help but wonder if . . .

Before you get too concerned, adjust the squelch or turn the knob to "test" on the "com" side of your radio. If you get the characteristic static, your radio is working. It's just that an unusual dearth of communication activity has occurred over the past few minutes.

When a previously-clear reception starts to break up, or you develop an unusual hum in the speaker, trouble may be brewing. Check the ammeter. Is it still showing a charge? Check the circuit breakers. If a navcom button has popped, let it cool off for a couple of minutes before resetting it. Does that do the trick? If so, you're probably all right . . . for a while. However, something is shorting out and should be checked as soon as you're back on the ground.

But suppose the ammeter shows no charge at all. Test it again by turning on the landing light. If the needle doesn't move, you can be sure the alternator has died or its belt has broken. In either case, what electrical power remains is coming from the battery alone. This being the case, turn off *all* nonessential electrical equipment, except one radio (assuming you can get some reception over it), and head for home or the nearest airport.

Unless you have experienced a radio or electrical failure in flight, the ammeter is probably one of your least monitored instruments. We suggest that it be permanently incorporated in the panel scanning process. The sooner a problem is caught, the better, because the life of a battery, once the alternator is gone, is about two hours. After that, you'll have no electrical power for lights, navcoms, transponder, gauges, and the like. The engine won't stop, because the ignition system is independent of the alternator-battery system—but a functioning engine, as critical as that is, is about *all* you'll have left.

WHEN A FAILURE IS CONFIRMED

There is no question. The failure is real. Perhaps you can hear the tower and other aircraft, but your efforts to transmit meet with total silence. Or the worst situation: neither your transmitter *nor* your receiver is working.

To explain what happens now, let's set up four of the many possible VFR situations (IFR procedures are more complex and are covered in the *AIM*).

Case One: You're flying VFR locally around an uncontrolled airport, or you're on a cross-country but you're not inside a TCA or ARSA. Because two-way radio-communication is not *required* in this case, there is no need to alert ATC with special transponder codes. Just squawk 1200, stay away from TCAs, ARSAs, and ATAs, and land at an uncontrolled field to get your radios checked. Listen to the CTAF for traffic advisories if your receiver still works.

Case Two: This time you've been using Center, Approach, Departure or Tower and have been squawking a discrete code.

If your receiver still works, but all of a sudden no one can hear you, set the transponder to 7700 for one minute—be sure to check the time—and then 7600 for 15 minutes. (If you're still in the air after that 15 minutes, repeat the sequence until you're on the ground.) Continue to monitor your last frequency. The controller will instantly know you've got a problem. After one minute, when you switch to 7600, the controller will know that the problem is radio failure.

If you're near your announced destination, the controller will advise either Approach or Tower (if one or the other exists) of your problem. At the appropriate time, you will be advised to switch to the Approach or tower frequency for further vectoring to the airport. Whatever the situation, the controller will ask you to identify your intentions through questions. You press the ident button when the answer is in the affirmative and do nothing if it is negative.

The same basic procedure applies when you're using Approach into a TCA or ARSA. The only difference is that, if the failure occurs before you enter the TCA or ARSA and you haven't been in contact with ATC, in all likelihood, Approach will direct you to an airport *underlying* the TCA or ARSA—not the primary airport. The reason? At these airports, two-way communications are essential, except in a serious emergency. Even though you can receive instructions, that's not enough. Also, if you've already got a bad transmitter, how long will the receiver hold out? Your radio is getting shaky, and this is no time to be fiddling around in a heavy traffic area. If you do lose the receiver after you've talked to Approach and are in a TCA, ARSA, or TRSA, continue on your intended course to the primary airport, and expect light signals from the tower (discussed later in this chapter).

Case Three: You have not used Center or Approach but experience the problem and want to land at a non-TCA, non-ARSA tower-controlled airport. Squawk 7700/7600, and if your receiver is working, tune to the tower frequency. If the tower is radar-equipped, the handling will be similar to that in the first or second example. If the doesn't have radar. What now?

Well, perhaps Center or Approach has picked up your code and has alerted the tower. But, in any event, keep a close eye out for other aircraft, and listen for the traffic pattern, instructions, and active runway. Then fly down that runway about 500 feet above the pattern altitude. You might have to repeat the maneuver a couple of times to attract attention, but an alert controller will probably spot you the first time. If he does, you'll hear something like: "Aircraft over the runway, Lincoln Tower, if you receive me, rock your wings." Followed by: "Roger. Do you intend to land at Lincoln?" Since that's your intention, you rock the wings again. The controller will then fit you into the pattern. At night, you might be asked to acknowledge by blinking your landing light or position lights.

If you have reason to believe that your transmitter works but your receiver doesn't, broadcast your requests over the tower frequency, and be alert for light signals from the tower.

If neither transmitter nor receiver is working, fly above the Airport Traffic Area to determine the direction and flow of traffic, then descend and join the traffic pattern, and watch for light signals. Acknowledge the light signals the same way as above.

Case Four: You are approaching a TCA or ARSA and want to land at the primary airport (or transit the airspace). You have not been in touch with Center and your radios fail *before* you are able to contact Approach. In this case, if you still want to enter the TCA or ARSA, you must land *outside* of it and request permission by telephone.

If your receiver works, you might try the 7700/7600 routine and monitor the appropriate Approach Control frequency for instructions. But if you get no response, you *must* avoid the TCA or ARSA, except in a true emergency.

THE TOWER LIGHT SYSTEM

Just to make this dicussion complete, we should remind you of the tower light system and what the different signals mean. There's nothing new about the system; it is a routine part of the private pilot training program. For those who may have forgotten the visual instructions the tower gives during two-way radio failure, the light gun signals are shown in TABLE 12-1.

Table 12-1. Tower Light Signals.

	MEANING		
COLOR AND TYPE OF SIGNAL	MOVEMENT OF VEHICLES EQUIPMENT AND PERSONNEL	AIRCRAFT ON THE GROUND	AIRCRAFT IN FLIGHT
Steady green	Cleared to cross, proceed, or go	Cleared for takeoff	Cleared to land
Flashing green	Not applicable	Cleared for taxi	Return for landing (to be followed by steady green at the proper time)
Steady red	STOP	STOP	Give way to other aircraft and continue circling
Flashing red	Clear the taxiway/runway	Taxi clear of the runway in use	Airport unsafe, do not land
Flashing white	Return to starting point on airport	Return to starting point on airport	Not applicable
Alternating red and green	Exercise extreme caution	Exercise extreme caution	Exercise extreme caution

CONCLUSION

The preceding recommendations, suggestions, procedures, or what have you, are fine; they will *almost* always get you down without any problems. But they don't work 100 percent of the time.

One of the authors once lost an alternator belt in the vicinity of a TCA and an underlying radar-equipped airport. We could receive, although the reception was scratchy and getting worse, but not transmit. Tuned to the tower and squawking 7700/7600 produced no response, so we flew up the landing runway at 2000 feet AGL.

Midway over the airport, we heard: "Aircraft over the runway, rock your wings if you receive me and are landing at _____."

Replying affirmatively, we were sequenced into the pattern and cleared to land.

Once off the runway, we switched to Ground, but if any taxi clearance was transmitted, the radio was too far gone for us to know. And there was no light from the tower.

After a couple of futile minutes, we taxied slowly to the ramp and placed a telephone call to the tower to explain our unauthorized taxiing and to ask if the 7700/7600 transmission had been received. The response was no—because the radar was down. Why no light signals? Sorry about that, but "our gun is broken."

So, not all malfunctions involve the aircraft. In this case, it was the alertness of the controller who spotted us above the runway and led us down to an uneventful landing at a busy airport.

Radio failures are relatively rare, but when they do occur, you should know what to do. Don't bust into a controlled airport with no radios unless you feel that you have no safe alternative. Get out of the vicinity, find an uncontrolled field, and have the problem repaired (or make a phone call to the tower to request a no-radios clearance).

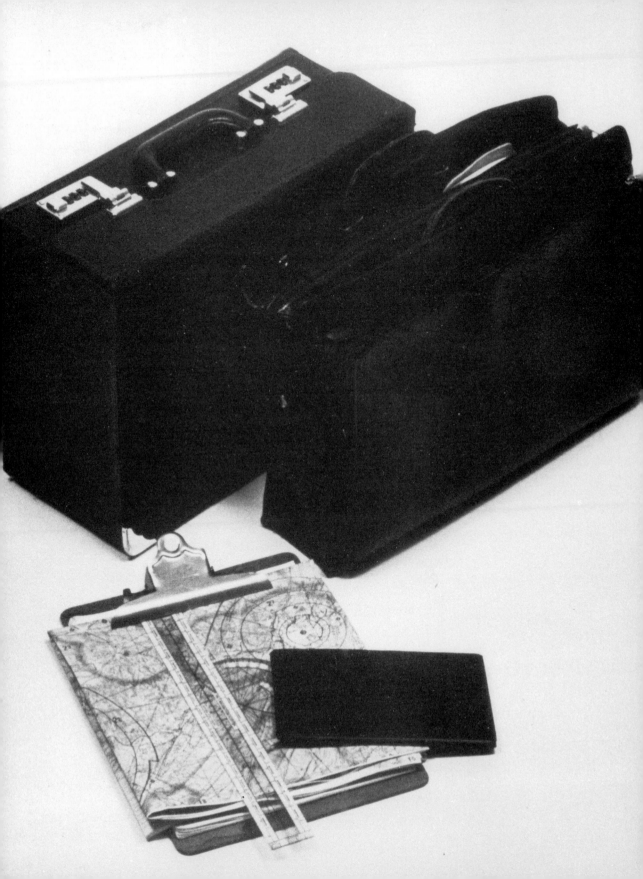

13

A Cross-Country: To Put It All Together

The preceeding chapters have taken us from MULTICOM through Center. Now let's put it all together with a mythical cross-country trip. Except for MULTICOM and TRSAs, we'll try to incorporate at least one example of the radio communications with each of the various elements we've been discussing. Hopefully, then, the "flight" will serve as a reasonable model for the real-life excursions you may make. Naturally, your itineraries will be different, as will the frequencies, but the principles illustrated should not vary because of that.

This flight will take us from Kansas City Downtown Airport to Omaha's Eppley Field, where we'll pick up two friends. From there, we'll go to Minneapolis International Airport. After completing our business in the Twin Cities, we'll drop one of our friends off at Mason City, Iowa, then the other friend at Newton, Iowa. From Newton, it's nonstop back to Kansas City.

The entire flight presupposes that you have a Mode C transponder and two navcoms. If you don't have the luxury of multiple navcoms the principles are the same, but frequency-changing is a little less convenient.

THE FLIGHT ROUTE

We decide to fly the airways whenever possible, even though doing so will add miles to the trip. Accordingly, the route of flight, VORs, course headings, and point-to-point nautical mileage resemble FIG. 13-1.

As 800-plus nautical miles is obviously too much territory to cover in one day, with stops and business en route, we plan to remain overnight in Minneapolis.

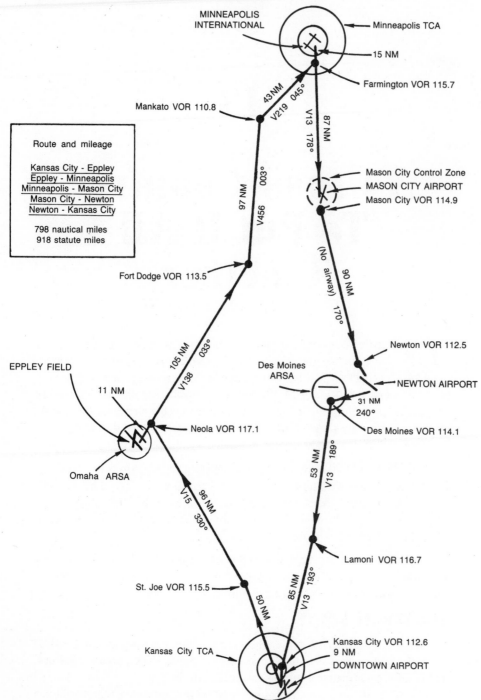

Fig. 13-1. *The route of the simulated cross-country.*

We'll also refuel at each stop, except Mason City, and use Center over each segment.

Why this itinerary? Simply to expose us to the various situations, airports, and control agencies that were discussed in the previous chapters. To wit:

✈ Kansas City Downtown lies *under* a TCA.

✈ Omaha Eppley has an ARSA.

✈ Minneapolis International is a TCA airport.

✈ Mason City has a Control Zone and a weather observer, but no tower or FSS on the field.

✈ Newton is an uncontrolled field with only UNICOM.

To go through the entire flight planning process is beyond the scope of this book. Let's assume that all preliminaries have been completed, including weight-and-balance calculations and filing the flight plan.

From here on it is no longer "us" and "we," but "you." You'll do the flying, make the calls, and be responsible for responding when you hear your N-number over the speaker or headset. But before going too far, a word about frequencies and getting them organized ahead of time.

RECORDING THE FREQUENCIES

As part of the preflight planning, write down *in sequence* the known or probable frequencies you will use. Some, particularly Center's, may differ from what you expect, but at least you'll be ready for the majority that will come into play.

As a suggestion, don't record all the frequencies on one piece of paper for a flight like this with five different legs. Enter the frequencies to Omaha on one sheet, those from Omaha to Minneapolis on another, and so on. (SEE FIGS. 13-2 THROUGH 13-6). Then number each sheet in sequence. As you leave one frequency and progress to the next, draw a line through (but don't obliterate) the one you have just left. This preliminary recording and progressive deleting will provide cockpit organization and minimize some of the confusion that is often the bane of the private pilot.

Two other suggestions: On each segment page, provide space for the ATIS information at the destination airport. When you're on the ground before departure, this isn't so important because you can listen to the local information as many times as necessary. In the air, however, it's another matter. Center has handed you off to Approach, but before contacting Approach, you should have monitored the ATIS (either over your second radio, or over the VOR frequency, if available)—which means that you don't have a lot of time to absorb the data being transmitted. Approach is expecting to hear from you rather promptly after the handoff.

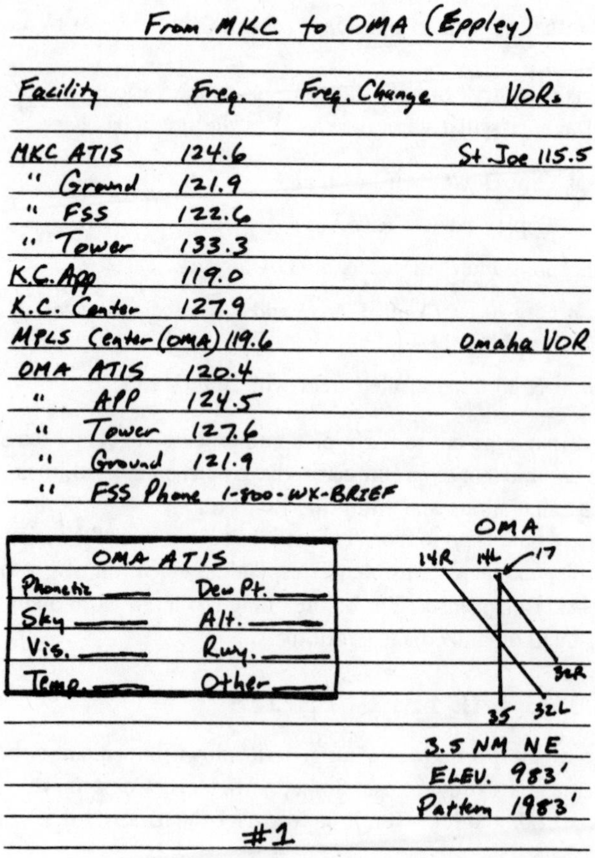

From MKC to OMA (Eppley)

Facility	Freq.	Freq. Change	VOR.
MKC ATIS	124.6		St. Joe 115.5
" Ground	121.9		
" FSS	122.6		
" Tower	133.3		
K.C. App	119.0		
K.C. Center	127.9		
MPLS Center (OMA)	119.6		Omaha VOR
OMA ATIS	120.4		
" APP	124.5		
" Tower	127.6		
" Ground	121.9		
" FSS Phone	1-800-WX-BRIEF		

OMA ATIS

Phonetic ___	Dew Pt. ___		
Sky ___	Alt. ___		
Vis. ___	Ruy. ___		
Temp. ___	Other ___		

OMA

14R 14L 17

32R

35 32L

3.5 NM NE
ELEV. 983'
Pattern 1983'

#1

Fig. 13-2. *Cross-country pre-flight planning notes: MKC to OMA (Eppley).*

Consequently, to expedite matters, line out a box on the flight segment page so that you can record the critical information: (1) the phonetic alphabet; (2) the sky or ceiling; (3) visibility; (4) temperature; (5) dewpoint; (6) altimeter setting; (7) runway in use; (8) other important information that might be included. Now, when you call Approach, you can advise the controller that you "have Charlie," or whatever, and be sure that you have it accurately. Memories do fail us.

Second, sketch in on the same segment page a rough diagram of the destination airport runways, including the distance and direction from town, field elevation, and pattern altitude. If you want, you can add the taxiways and building locations. *AOPA'S Airports USA*, published by the Aircraft Owners and Pilots Association, provides diagrams and data of over 13,000 airports in the U.S. and its possessions. It's an excellent source for determining the layout and runway data of whatever airport you have in mind. Similar diagrams are found on state aeronautical charts and on instrument approach charts.

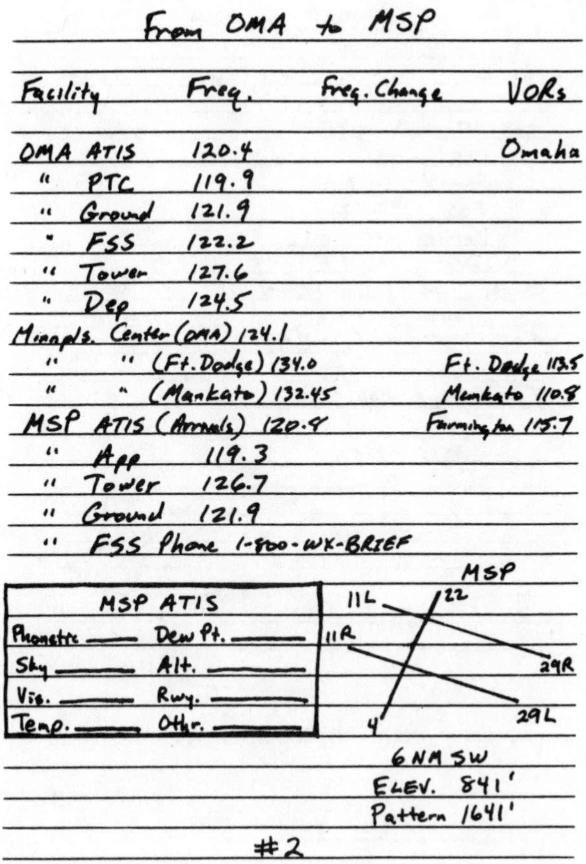

From OMA to MSP

Facility	Freq.	Freq. Change	VORs
OMA ATIS	120.4		Omaha
" PTC	119.9		
" Ground	121.9		
" FSS	122.2		
" Tower	127.6		
" Dep	124.5		
Mnnpls. Center (OMA)	124.1		
" " (Ft. Dodge)	134.0		Ft. Dodge 113.5
" " (Mankato)	132.45		Mankato 110.8
MSP ATIS (Arrivals)	120.8		Farmington 115.7
" App	119.3		
" Tower	126.7		
" Ground	121.9		
" FSS Phone	1-800-WX-BRIEF		

```
+---------------------------------+          MSP
|          MSP ATIS        |  11L    / 22
+---------------------------------+
| Phonetic ___  Dew Pt. ___| 11R
+---------------------------------+                    29R
| Sky ___       Alt. ___   |
+---------------------------------+
| Vis. ___      Rwy. ___   |
+---------------------------------+
| Temp. ___     Other ___  |  4'              29L
+---------------------------------+
                        6 NM SW
                        ELEV. 841'
                        Pattern 1641'
                  #2
```

Fig. 13-3. *Cross-country pre-flight planning notes: EPP to MPS.*

The purpose of the sketch is apparent: it minimizes mental or spatial confusion when going into a strange airport for the first time. Just as important, it can reduce the number of questions or inquiries you might have to make of Approach or Tower.

The examples (FIGS. 13-2 THROUGH 13-6) aren't very fancy, but that's not the point. Practicality is the objective. (Just don't rely on the frequencies cited as being current. They do change.)

THE FLIGHT AND THE RADIO CONTACTS

Equipped with the necessary charts—Sectionals as well as the Enroute Low Altitudes—the frequencies recorded, the flight planned, and the flight plan filed, you're ready to depart.

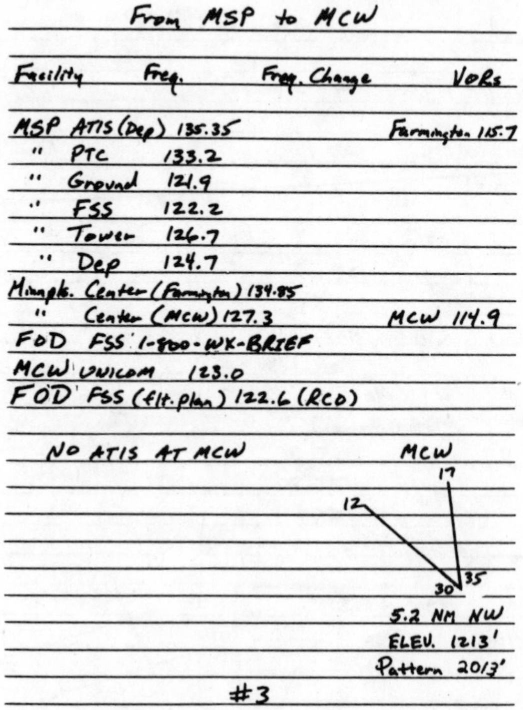

From MSP to MCW

Facility	Freq.	Freq. Change	VORs
MSP ATIS (Dep)	135.35		Farmington 115.7
" PTC	133.2		
" Ground	121.9		
" FSS	122.2		
" Tower	126.7		
" Dep	124.7		
Minnpls. Center (Farmington)	134.85		
" Center (MCW)	127.3		MCW 114.9
FOD FSS	1-800-WX-BRIEF		
MCW UNICOM	123.0		
FOD FSS (flt. plan)	122.6 (RCO)		

NO ATIS AT MCW

MCW
17
12
30 35
5.2 NM NW
ELEV. 1213'
Pattern 2013'

#3

Fig. 13-4. *Cross-country pre-flight planning notes: MSP to MCW.*

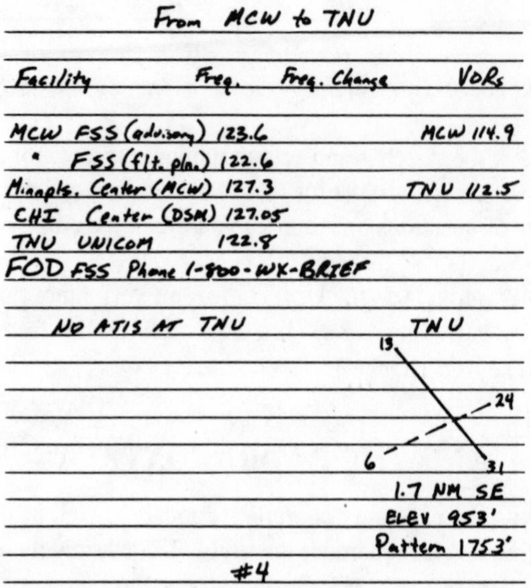

From MCW to TNU

Facility	Freq.	Freq. Change	VORs
MCW FSS (advisory)	123.6		MCW 114.9
" FSS (flt. plan.)	122.6		
Minnpls. Center (MCW)	127.3		TNU 112.5
CHI Center (DSM)	127.05		
TNU UNICOM	122.8		
FOD FSS Phone	1-800-WX-BRIEF		

NO ATIS AT TNU

TNU
13
24
6
31
1.7 NM SE
ELEV 953'
Pattern 1753'

#4

Fig. 13-5. *Cross-country pre-flight planning notes: MCW to TNU.*

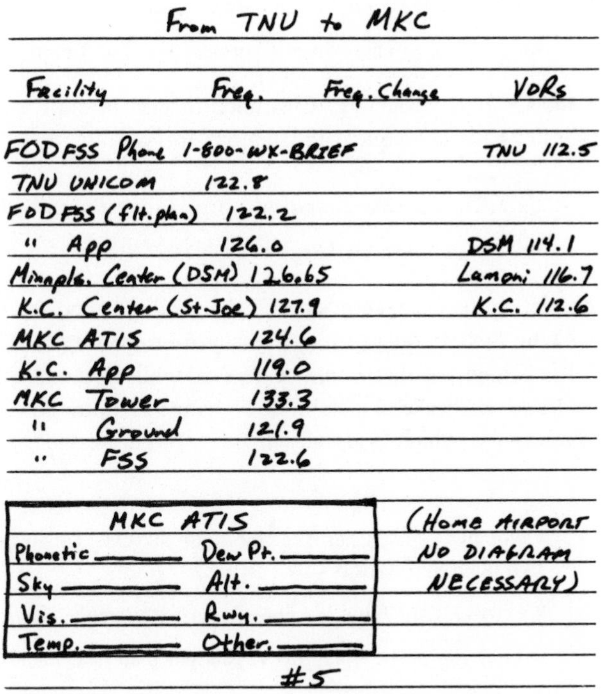

From TNU to MKC

Facility	Freq.	Freq. Change	VORs
FOD FSS Phone	1-800-WX-BRIEF		TNU 112.5
TNU UNICOM	122.8		
FOD FSS (flt. plan)	122.2		
" App	126.0		DSM 114.1
Minnpls. Center (DSM)	126.65		Lamoni 116.7
K.C. Center (St. Joe)	127.9		K.C. 112.6
MKC ATIS	124.6		
K.C. App	119.0		
MKC Tower	133.3		
" Ground	121.9		
" FSS	122.6		

MKC ATIS		(HOME AIRPORT
Phonetic _____ Dew Pt. _____		NO DIAGRAM
Sky _____ Alt. _____		NECESSARY)
Vis. _____ Rwy. _____		
Temp. _____ Other. _____		

#5

Fig. 13-6. *Cross-country pre-flight planning notes: TNU to MKC.*

Kansas City to Omaha

With the engine started, the first order of business is to tune to the Downtown Airport ATIS on 124.6. Put this in the #1 radio and set up Ground Control, 121.9, on #2.

ATIS: *This is Kansas City Downtown Airport Information Delta. One six four five Zulu weather. Five thousand scattered, measured ceiling ten thousand broken, visibility eight. Temperature seven eight, dewpoint five five, wind one seven zero at ten, altimeter two niner niner eight. ILS Runway One Niner in use, land and depart Runway One Niner. Advise you have Delta.*

Now change #1 to the FSS frequency of 122.6, which is the frequency Flight Service gave you to open your flight plan. Put the transponder on STANDBY and call Ground Control on #2 radio:

You: Downtown Ground, Cherokee One Four Six One Tango at the Flying Service, VFR Omaha with Information Delta.

GC: *Cherokee One Four Six One Tango, taxi to Runway One Niner.*

You: Roger. Taxi to One Niner, Cherokee Six One Tango.

Stay on the Ground frequency. Clearance to taxi doesn't mean that the controller may not have subsequent instructions for you:

GC: *Cherokee Six One Tango, give way to the Baron taxiing south.*

You: Wilco, Cherokee Six One Tango.

After completing the pre-takeoff check, either call Ground and advise them that you're leaving their frequency momentarily to go to Flight Service, or, if there's no traffic behind you that would be delayed while you make that contact, taxi to the hold line and brake to a stop. Now change #2 to the tower frequency of 133.3, switch to #1, which is already tuned to Flight Service, and open the flight plan:

You: Columbia Radio, Cherokee One Four Six One Tango on one two two point six, Kansas City.

FSS: *Cherokee One Four Six One Tango, Columbia Radio.*

You: Would you please open my VFR flight plan to Omaha Eppley at this time?

FSS: *Cherokee Six One Tango, roger. We'll open your flight plan at five five.*

You: Roger, thank you. Cherokee Six One Tango.

Two points to remember about this:

✈ Be sure to add five minutes, even ten, to your flight plan arrival time in case your departure is delayed.

✈ Opening a flight plan while still on the ground is possible only when there is an FSS, RCO, or VOR voice facility on the field.

As your course to Omaha is northwesterly, the most direct routing is through the Kansas City TCA, so change #1 radio to Approach Control on 119.0. (Remember that you're under the TCA and must contact Approach to enter it.) With #1 set up, the next call is to the tower on #2:

You: Downtown Tower, Cherokee One Four Six One Tango ready for takeoff with north departure.

Twr: *Cherokee One Four Six One Tango, hold short. Landing traffic.*

You: Cherokee Six One Tango holding short.

Twr: *Cherokee Six One Tango, cleared for takeoff. Left turn for north departure approved. Remain clear of the final approach course. Contact Approach when airborne.*

You: Will do. Cherokee Six One Tango. [Now switch the transponder to ALT.]

Even if the tower has advised you to contact Approach after takeoff, it's wise to request the frequency-change approval or to inform the tower that you're about to make the change. The controller might have reasons for wanting you to stay with him for a few minutes. Regardless, the next call would be:

You: Tower, Cherokee Six One Tango requests frequency change [or "going to Approach"].

Twr: *Cherokee Six One Tango. Frequency change approved.*

You: Cherokee Six One Tango. Good day.

Go now to the #1 radio and call Approach on 119.0:

You: Kansas City Approach, Cherokee One Four Six One Tango.

App: *Cherokee One Four Six One Tango, Kansas City Approach, go ahead.*

You: Cherokee One Four Six One Tango is off Downtown at two thousand, requesting six thousand five hundred to Omaha, and would like clearance through the TCA.

App: *Cherokee Six One Tango, squawk zero two five two and ident. Remain clear of the TCA until radar contact.*

You: Cherokee Six One Tango squawking zero two five two.

Remember that you have not yet been cleared into the TCA, so stay below the 2400-foot floor until you hear the next message:

App: *Cherokee Six One Tango, radar contact. Cleared through the TCA. Fly heading three four zero and maintain four thousand five hundred.*

You: Roger. Cleared through the TCA, heading three four zero, leaving two thousand for four thousand five hundred. Cherokee Six One Tango.

As you start your turn and begin the climb, Approach may come on again:

App: *Cherokee Six One Tango, traffic at twelve o'clock, three miles, southbound. Unverified altitude two thousand niner hundred.*

You: Cherokee Six One Tango is looking.

And then:

> **App:** *Cherokee Six One Tango, traffic no longer a factor.*
>
> **You:** Roger. Cherokee Six One Tango.
>
> **You:** Cherokee Six One Tango, level at four thousand five hundred.
>
> **App:** *Cherokee Six One Tango, roger. Climb and maintain six thousand five hundred.*
>
> **You:** Out of four thousand five hundred for six thousand five hundred. Cherokee Six One Tango.
>
> **You:** Cherokee Six One Tango level at six thousand five hundred.
>
> **App:** *Cherokee Six One Tango, roger. Turn left heading three one zero. Proceed direct St. Joe when able.*
>
> **You:** Roger. Three one zero on the heading. Direct St. Joe when able. Cherokee Six One Tango.

The St. Joe VOR lies about 50 NM north of Kansas City. Approach is saying that you are to tune the nav receiver to the VOR frequency of 115.5. When you have the altitude and distance to get a steady needle reading, you're cleared to proceed directly on course to St. Joe.

Other instructions or traffic alerts may follow. While you have the time, however, you should be setting up the #2 radio (until now still on the tower frequency) to Kansas City Center, which you expect to be 127.9. Approach might give you a different frequency, but if not, you're ready to contact Center without delay.

As you work your way northward toward St. Joe, you reach the TCA limits:

> **App:** *Cherokee Six One Tango, position 20 miles northwest of International, departing the TCA. Radar service terminated. Squawk one two zero zero. Frequency change approved.*
>
> **You:** Cherokee Six One Tango. Can you hand us off to Center?
>
> **App:** *Unable at this time, Six One Tango. Contact Center on one two seven point niner. Good day.*
>
> **You:** One two zero zero and one two seven point niner. Roger. Cherokee Six One Tango.

There is no handoff, so the call to Center requires the full IPAIDS:

> **You:** Kansas City Center, Cherokee One Four Six One Tango.
>
> **Ctr:** *Cherokee One Four Six One Tango, Kansas City Center, go ahead.*

You:	Cherokee One Four Six One Tango is 20 south of St. Joe VOR, level at six thousand five hundred en route Omaha. Squawking one two zero zero. Request VFR advisories, if possible.
Ctr:	*Cherokee Six One Tango, squawk four one five zero and ident.*
You:	Cherokee Six One Tango squawking four one five zero.
Ctr:	*Cherokee Six One Tango. Radar contact. Altimeter two niner niner eight.*
You:	Roger. Cherokee Six One Tango.

What comes over the air now will depend on traffic in your line of flight of which you should be aware. You may be told to assume different headings; you may be alerted to the proximity of other aircraft; you may be warned of potential military training exercises; you may hear nothing—until:

Ctr:	*Cherokee Six One Tango, contact Minneapolis Center now on one one niner point six. Good day.*
You:	One one niner point six. Thank you for your help. Cherokee Six One Tango.

No comment here about radar service being terminated, so this is a handoff by Kansas City to Minneapolis. If you've already changed the #1 radio from Kansas City Approach to Minneapolis, all you have to do now is go back to #1 and introduce yourself:

You:	Minneapolis Center, Cherokee One Four Six One Tango with you, level at six thousand five hundred.
Ctr:	*Cherokee Six One Tango, roger. Altimeter three zero zero two.*

Unless you are told otherwise, maintain your present heading and altitude. Again, there may or may not be instructions or advice for you. Eventually, however, as you near the Omaha ARSA, Center will come back on:

Ctr:	*Cherokee Six One Tango, position 30 miles south of Eppley. Contact Omaha Approach on one two four point five. Good day.*
You:	Roger. One two four point five. Cherokee Six One Tango. Good day.

This, too, is a handoff, so tune in to Approach. Before making the call, however, if you haven't done so already, get the ATIS on 120.4 for the current Eppley information. Now you're ready to contact Approach:

> You: Omaha Approach, Cherokee One Four Six One Tango is with you, level at six thousand five hundred with Information Echo.
>
> *App:* *Cherokee Six One Tango, roger.*

As you enter the ARSA and draw close to the field, you're likely to be given a variety of instructions that will sequence you into the existing traffic flow. Whatever the messages, be sure to acknowledge and repeat them in an abbreviated form:

> *App:* *Cherokee Six One Tango, turn right heading three five zero. Descend and maintain four thousand.*
>
> You: Right to three five zero. Leaving six thousand five hundred for four thousand. Cherokee Six One Tango.

In another few minutes, you'll hear from Approach again:

> *App:* *Cherokee Six One Tango, Eppley is at twelve o'clock, six miles. Contact Eppley Tower on one two seven point six.*
>
> You: Roger. One two seven point six—and we have the field in sight. Cherokee Six One Tango. Good day.

Another handoff:

> You: Eppley Tower, Cherokee One Four Six One Tango with you, level at four thousand.
>
> *Twr:* *Cherokee Six One Tango, enter left downwind for Runway One Four. Sequence later.*

Two points here:

- "Sequence later" simply means the tower will advise you in due time whether you're cleared to land, are "number two following a Cessna on downwind," "number four following the Duchess," or whatever. Just remember to tell the tower that you "have the Cessna" or "negative contact on the Duchess" or "have the traffic." Always keep the tower informed—don't leave them in the dark.

- You'll note that you've used the same squawk from Kansas City Center through Minneapolis Center, Omaha Approach, and Eppley Tower. No controlling agency has asked you to change. This is not always the case. Any one of them could have requested a different squawk and an ident. Leave the transponder on the current squawk until directed otherwise.

With a little time available now, dial in Eppley's Ground frequency, 121.9, on #1 Radio. Then, in due course, you'll hear on #2:

Twr: *Cherokee Six One Tango, cleared to land.*

You: Roger, cleared to land. Cherokee Six One Tango.

During the landing roll-out, the tower makes its final contact with you:

Twr: *Cherokee Six One Tango, contact Ground point niner clear of the runway.*

You: Cherokee Six One Tango, wilco.

Suppose, however, that you're not sure whether to make a left or right turn-off. You want to go to a Texaco dealer but don't know where one is located. Despite the uncertainty, don't tie up the tower frequency by asking the controller for directions. Let Ground do this for you, even if it means being cleared back across the active runway because you turned right instead of left, or vice versa. Merely acknowledge the tower's instructions and go to Ground's frequency:

You: Eppley Ground, Cherokee One Four Six One Tango clear of the active. Request progressive taxi to the Texaco dealer.

GC: *Cherokee Six One Tango, continue straight ahead to the second intersection and hold.*

You: Second intersection and hold. Cherokee Six One Tango.

GC: *Cherokee Six One Tango. Turn right onto Taxiway Echo. Texaco sign will be on your left. Good day.*

You: Roger. Right onto Echo. Texaco on the left. Thank you for your help. Cherokee Six One Tango.

Now don't forget to call Flight Service either by radio or phone to close out your flight plan!

Omaha to Minneapolis

After an hour on the ground for refueling, a bite to eat, and a call to the Columbus, Nebraska, AFSS for a weather briefing and filing the flight plan, you're ready to go again, with friends and baggage on board. The only new element will be the need to call Clearance Delivery for initial VFR instructions within the ARSA. As you've already determined, the Pre-Taxi Clearance frequency is 119.9, and you were told to contact Flight Service on the Omaha RCO frequency of 122.35. That's a change from the 122.2 you had listed on the frequency chart you prepared back in Kansas City (FIG. 13-3). So cross out 122.2 on that chart and enter 122.35 under the "Changes" column.

As usual, the communication chain begins by monitoring ATIS: *"This is Eppley Information Xray. One niner four five Zulu weather. Eight thousand scattered, visibility five, haze and smoke. Temperature eight five, dewpoint six two. Wind one four zero at one five. Altimeter three zero two five. ILS Runway One Four Right in use, land and depart Runway One Four Right. Advise you have Gulf."*

Next, the call to Clearance:

You:	Eppley Clearance, Cherokee One Four Six One Tango.
PTC:	*Cherokee One Four Six One Tango, Eppley Clearance.*
You:	Cherokee One Four Six One Tango will be departing Eppley, VFR northeast for Minneapolis at seven thousand five hundred.
PTC:	*Cherokee Six One Tango, roger. Turn left heading zero four five after departure. Climb and maintain three thousand. Squawk two four four zero. Departure frequency one two four point five.*
You:	Roger. Zero four five on the heading, maintain three thousand, two four four zero, and one two four point five. Cherokee Six One Tango.
PTC:	*Cherokee Six One Tango, readback correct.*
You:	Cherokee Six One Tango.

Before calling Ground on the #2 radio, dial out the Clearance frequency in #1 and replace it with Flight Service's—122.35. Also, put the transponder on STANDBY.

You:	Eppley Ground, Cherokee One Four Six One Tango at Texaco with Information Gulf. Ready to taxi with clearance.
GC:	*Cherokee Six One Tango, taxi to Runway One Four Right for intersection departure.*
You:	Ground, Cherokee Six One Tango would like full length.
GC:	*Cherokee Six One Tango, roger. Full length approved.*
You:	Cherokee Six One Tango.

The pre-takeoff check completed, the next call is to Flight Service after advising Ground that you were going to change frequencies momentarily.

You:	Columbus Radio, Cherokee One Four Six One Tango on one two two point three five.
FSS:	*Cherokee One Four Six One Tango, Columbus Radio.*
You:	Columbus, Cherokee Six One Tango. Would you open my VFR flight plan to Minneapolis International at this time?
FSS:	*Cherokee Six One Tango, roger. We'll open your flight plan at two five.*
You:	Roger, thank you. Cherokee Six One Tango.

That done, the next move is to taxi to the hold line and call the tower:

You:	Eppley Tower, Cherokee One Four Six One Tango ready for take-off with northeast departure.
Twr:	*Cherokee Six One Tango, taxi into position and hold.*
You:	Position and hold. Cherokee Six One Tango. [Switch the transponder from STANDBY to ALT as you're taxiing to the runway.]
Twr:	*Cherokee Six One Tango, cleared for takeoff.*
You:	Roger, cleared for takeoff. Cherokee Six One Tango. Is northeast departure approved?
Twr:	*Affirmative, northeast departure approved.*
You:	Roger. Cherokee Six One Tango.

When airborne, request the frequency change to Departure—if the tower has not already advised you to do so:

You:	Tower, Cherokee Six One Tango requests frequency change to Departure.
Twr:	*Cherokee Six One Tango, frequency change approved.*
You:	Roger. Cherokee Six One Tango. Good day.

To Departure Control:

You:	Omaha Departure, Cherokee One Four Six One Tango is with you out of one thousand eight hundred for three thousand. Request seven thousand five hundred.
Dep:	*Cherokee Six One Tango, radar contact. Maintain present heading. Report level at three thousand.*
You:	Roger, report three thousand, Cherokee Six One Tango.

You:	Cherokee Six One Tango level at three thousand.
Dep:	*Cherokee Six One Tango, roger. Turn left heading zero three zero.*
You:	Left to zero three zero. Cherokee Six One Tango.
Dep:	*Cherokee Six One Tango, climb and maintain seven thousand five hundred.*
You:	Roger. Out of three thousand for seven thousand five hundred. Cherokee Six One Tango.
You:	Cherokee Six One Tango level at seven thousand five hundred.
Dep:	*Cherokee Six One Tango, roger. Contact Minneapolis Center on one two four point one.* [An unsolicited handoff for advisories]
You:	One two four point one. Thank you for your help. Cherokee Six One Tango.

To Center:

You:	Minneapolis Center, Cherokee One Four Six One Tango with you, level at seven thousand five hundred.
Ctr:	*Cherokee Six One Tango, roger. Radar contact.*
You:	Roger. Cherokee Six One Tango.

Down the road apiece, the summer turbulence at 7500 feet is getting a bit too much for one of your passengers, so you decide to climb to 9500. But first:

You:	Center, Cherokee Six One Tango is out of seven thousand five hundred for niner thousand five hundred due to turbulence.
Ctr:	*Cherokee Six One Tango, roger. Report level at niner thousand five hundred.*
You:	Wilco, Cherokee Six One Tango.
You:	Cherokee Six One Tango level at niner thousand five hundred.
Ctr:	*Cherokee Six One Tango, roger.*

Just remember that on a VFR flight plan outside of a terminal area, you're free to deviate from existing headings and altitudes. But since you're asking Center for en route advisories, don't make changes without advising the controller of your intentions. Although he can spot heading changes on the screen (and altitude changes, if you're equipped with a Mode C transponder), be sure to warn him in advance.

As you proceed toward Fort Dodge on V138, Center comes on again:

Ctr: *Cherokee Six One Tango, contact Minneapolis Center now on one three four point zero.* [This is the Fort Dodge remote outlet.]

You: Roger. One three four point zero. Cherokee Six One Tango. Good day.

Change the frequency accordingly and reestablish yourself with Center:

You: Minneapolis Center, Cherokee One Four Six One Tango with you level at niner thousand five hundred.

Ctr: *Cherokee Six One Tango, roger. Altimeter three zero one six.* [Plus any instructions or traffic advisories the controller might have for you.]

Crossing the Fort Dodge VOR, you make a time check against your flight plan ETA. What the check reveals is that you're running about 20 minutes behind schedule, which is quite a difference for a 105-mile flight. The forecast winds must have changed or the new altitude of 9500 feet is producing winds from a different direction and/or velocity. Whatever the case, and because you intend to stay at 9500 and fly in a generally north-northeast direction, it's fair to assume that your estimated time to Minneapolis will be slower than expected. You may not be 30 or more minutes late, but why take a chance? With another 150 miles to go, it wouldn't take much to extend the flight 30 or 45 minutes.

Before taking any arbitrary action, you decide that more information about the winds is in order, thus a call to Flight Service. First, however, advise Center if you're temporarily going to leave the frequency:

You: Center, Cherokee Six One Tango is leaving you temporarily to go to Flight Service.

Ctr: *Cherokee Six One Tango, roger. Advise when you're back with me.*

You: Will do. Cherokee Six One Tango.

The FSS call is made on 122.3, the Fort Dodge transmit/receive frequency:

You: Fort Dodge Radio, Cherokee One Four Six One Tango on one two two point three.

FSS: *Cherokee One Four Six One Tango, Fort Dodge Radio, go ahead.*

You: Fort Dodge, Cherokee Six One Tango is just north of the Fort Dodge VOR at niner thousand five hundred. Request winds at niner thousand feet.

> *FSS:*   *Cherokee Six One Tango, roger. Stand by.*
>
> *FSS:*   *Cherokee Six One Tango, winds at niner thousand are three five zero at four zero. Fort Dodge altimeter two niner two five.*
>
> You: Three five zero at four zero and two niner two five. Thank you. Cherokee Six One Tango.

Without going through the mechanics of recomputing your ground speed and ETA based on this information, let's just say that you determine that your arrival will be 48 minutes later than your flight plan forecast. This conclusion warrants another call to Flight Service:

> You: Fort Dodge Radio, Cherokee One Four Six One Tango on one two two point three.
>
> *FSS:*   *Cherokee One Four Six One Tango, Fort Dodge Radio, go ahead.*
>
> You: Cherokee One Four Six One Tango is 10 north of the Fort Dodge VOR on VFR flight plan to Minneapolis International with a one six four five local ETA. Would like to extend the ETA to one seven one five.
>
> *FSS:*   *Cherokee Six One Tango, roger. Will extend your ETA to one seven one five local. Fort Dodge altimeter two niner two five.*
>
> You: Roger, thank you. Cherokee Six One Tango.

The next move is to go back to Center and reestablish contact:

> You: Center, Cherokee Six One Tango is back with you.
>
> *Ctr:*   *Cherokee Six One Tango, roger.*

With that taken care of, you can rest more easily. You have until 1745 before Flight Service will start asking questions as to your whereabouts. You've also allowed yourself an additional twelve minutes as an extra cushion.

Moving northward along V456, you'll be approaching another Center change point—this time to the remoted Mankato site. Somewhere near Mankato, you get this call:

> *Ctr:*   *Cherokee Six One Tango, contact Minneapolis Center now on one three two point four five.*
>
> You: One three two point four five. Cherokee Six One Tango. Good day.
>
> You: Minneapolis Center, Cherokee One Four Six One Tango with you, level at niner thousand five hundred.

Ctr: *Cherokee Six One Tango, roger. Mankato altimeter two niner two three.*

You: Roger. Cherokee Six One Tango.

The closer you get to the Farmington VOR, the greater the likelihood of alerts, advisories, direction changes, and so forth. There may be none, but moving towards a busy VOR and TCA increases the probability.

Ten miles or so out of the TCA limits, Center will conclude its radar surveillance. Perhaps the controller will hand you off to Approach. If so, the usual "with you" is all that's necessary. Without handoff, you'll need to know your location and/or approximate distance from the airport. For a change, let's assume that there is no handoff:

Ctr: *Cherokee Six One Tango, radar service terminated. Squawk one two zero zero. Contact Minneapolis Approach on one one niner point three.*

You: One two zero zero and one one niner point three. Thank you for your help. Cherokee Six One Tango.

Before calling Approach, check International's arrival ATIS on 120.8 for the current arrival information. Then go to 119.3 for the initial contact with Approach:

You: Minneapolis Approach, Cherokee One Four Six One Tango.

App: *Cherokee One Four Six One Tango, Minneapolis Approach.*

You: Cherokee One Four Six One Tango is just east of Montgomery, level at niner thousand five hundred, landing International. Squawking one two zero zero with Information Echo.

App: *Cherokee Six One Tango, roger. Squawk four one two zero and ident. Remain outside the TCA until radar contact.*

You: Cherokee Six One Tango, squawking four one two zero, clear of the TCA.

App: *Cherokee Six One Tango, radar contact. Cleared into the Minneapolis TCA. Turn left heading zero three zero. Descend and maintain five thousand.*

You: Left to zero three zero. Out of niner thousand five hundred for five thousand. Cherokee Six One Tango.

You: Cherokee Six One Tango level at five thousand.

App: *Cherokee Six One Tango, roger. Proceed on course.*

App: *Cherokee Six One Tango, descend and maintain three thousand five hundred.*

You:	Out of five thousand for three thousand five hundred. Cherokee Six One Tango.
You:	Cherokee Six One Tango level at three thousand five hundred.
App:	*Cherokee Six One Tango, roger. Airport is at twelve o'clock, ten miles. Report in sight.*
App:	*Cherokee Six One Tango, TRAFFIC ALERT. Turn right heading zero seven zero IMMEDIATELY* [or "NOW"].

This was obviously a warning of nearby traffic, so you turn *now* to the new heading. Don't delay! When the controller—*any* controller—says "immediately" or "now," he means *now*—not a leisurely maneuver. No acknowledgment of the instruction is necessary.

Then, when the "crisis" is past:

App:	Cherokee Six One Tango, traffic no longer a factor. Turn left heading zero two five. Descend and maintain two thousand five hundred. Contact Minneapolis Tower on one two six point seven.
You:	Roger. Left to zero two five. Out of three thousand five hundred for two thousand five hundred. Cherokee Six One Tango. Good day.
You:	Minneapolis Tower, Cherokee One Four Six One Tango with you, out of three thousand five hundred for two thousand five hundred.
Twr:	*Cherokee Six One Tango, enter left base for Runway Two Niner Left.*
You:	Roger. Left base for Two Niner Left. Cherokee Six One Tango.

Unless further instructions are given, the next message will be:

Twr:	*Cherokee Six One Tango, cleared to land, Runway Two Niner Left.*
You:	Roger, cleared to land. Cherokee Six One Tango.

When you're down:

Twr:	*Cherokee Six One Tango, contact Ground point niner.*
You:	Roger, Cherokee Six One Tango.

When clear of the active and at a full stop:

You:	Minneapolis Ground, Cherokee One Four Six One Tango clear of Two Niner Left. Taxi to Airmotive.

GC: *Cherokee Six One Tango, taxi to Airmotive.*

Once again, when you're parked or in the pilot's lounge, be sure to close out your flight plan with Flight Service, which, in the case of Minneapolis, is located in Princeton, Minnesota.

Minneapolis to Mason City

It's the next morning and you're ready to set out for Mason City, Newton, and then back to Kansas City. First the normal routines: checking the weather, filing the flight plan, and determining the FSS frequency to open the flight plan. With the engine started, monitor the Minneapolis departure ATIS on 135.35. Remember that you're in a TCA, so the first call goes to Pre-Taxi Clearance on 133.2:

You: Minneapolis Clearance, Cherokee One Four Six One Tango.

PTC: *Cherokee One Four Six One Tango, Clearance.*

You: Cherokee One Four Six One Tango will be departing VFR to Mason City. Request seven thousand five hundred.

PTC: *Cherokee Six One Tango, roger. Turn right heading one seven five after departure. Climb and maintain three thousand. Squawk zero four two zero. Departure frequency one two four point seven.*

You: Right heading one seven five after departure, maintain three thousand, zero four two zero, and one two four point seven.

PTC: *Cherokee Six One Tango, readback correct.*

You: Roger. Cherokee Six One Tango.

Now to Ground Control on 121.9:

You: Minneapolis Ground, Cherokee One Four Six One Tango at Airmotive with Information Echo and clearance.

GC: *Cherokee Six One Tango, taxi to Runway One One Right.*

You: Roger, One One Right, Cherokee Six One Tango.

The pre-takeoff check has been completed. Before taxiing to the hold line, call Ground for permission to switch frequencies, then call Flight Service to open the flight plan.

You: Princeton Radio, Cherokee One Four Six One Tango on one two two point five five [the RCO frequency FSS gave you when filing], Minneapolis.

FSS:	*Cherokee One Four Six Tango, Princeton Radio.*
You:	Roger, would you please open my flight plan to Mason City at this time?
FSS:	*Cherokee Six One Tango, roger. Will open your flight plan at one zero.*
You:	Thank you. Cherokee Six One Tango.

As you move toward the hold line, you see that two aircraft are ahead of you awaiting takeoff permission. Regardless, you pull behind the second plane and call the tower on 126.7:

You:	Minneapolis Tower, Cherokee One Four Six One Tango ready for takeoff, number three in sequence, south departure.
Twr:	*Cherokee One Four Six One Tango, taxi around the Mooney and Skymaster. Cleared for takeoff, south departure approved.*
You:	Roger, cleared for takeoff, Cherokee Six One Tango.

At this point, switch the transponder from STANDBY to ALT. When airborne, turn to your assigned heading of 175°. The tower will probably authorize the frequency change to Departure, but if it doesn't, request the change:

You:	Tower, Cherokee Six One Tango requests frequency change to Departure.
Twr:	*Cherokee Six One Tango, frequency change approved. Good day.*
You:	Roger. Cherokee Six One Tango. Good day.

Switch to your other radio, already dialed in to 124.7:

You:	Minneapolis Departure, Cherokee One Four Six One Tango with you out of one thousand seven hundred for three thousand.
Dep:	*Cherokee One Four Six One Tango, report level at three thousand.*
You:	Wilco, Cherokee Six One Tango.
You:	Cherokee Six One Tango level at three thousand.
Dep:	*Cherokee Six One Tango, roger.*
Dep:	*Cherokee Six One Tango, climb and maintain four thousand five hundred.*
You:	Roger. Out of three thousand for four thousand five hundred. Cherokee Six One Tango.

You:	Cherokee Six One Tango level at four thousand five hundred.
Dep:	*Cherokee Six One Tango, roger. Cleared direct Farmington VOR.*
You:	Roger, cleared direct Farmington. Cherokee Six One Tango.
Dep:	*Cherokee Six One Tango, climb and maintain seven thousand five hundred.*
You:	Roger. Out of four thousand five hundred for seven thousand five hundred. Cherokee Six One Tango.
You:	Cherokee Six One Tango level at seven thousand five hundred.
Dep:	*Cherokee Six One Tango. Roger. Report Farmington VOR.*
You:	Roger, report Farmington. Cherokee Six One Tango.

It isn't long before the Course Direction Indicator (CDI) swings erratically and the VOR flag changes from TO to FROM. You're over the Farmington VOR. You report in and are cleared to turn to the 178-degree heading which establishes you on Victor 13. As you're still in the TCA, other instructions may be forthcoming from Departure. In a very few minutes, you'll hear something like this:

Dep:	*Cherokee Six One Tango, position five miles south of Farmington VOR, departing the TCA. Squawk one two zero zero. Radar service terminated. Frequency change approved. Good day.*
You:	Departure, Cherokee Six One Tango. Can you hand us off to Center?
Dep:	*Cherokee Six One Tango, stand by.* [Pause] *Cherokee Six One Tango, unable at this time. Squawk one two zero zero. Suggest you contact Minneapolis Center on one three four point eight five.*
You:	Roger. One two zero zero and one three four point eight five. Cherokee Six One Tango.
You:	Minneapolis Center, Cherokee One Four Six One Tango.
Ctr:	*Cherokee One Four Six One Tango, Minneapolis Center.*
You:	Cherokee One Four Six One Tango is five south of the Farmington VOR at seven thousand five hundred, VFR to Mason City, squawking one two zero zero. Request VFR advisories, if possible.

This time, the press of traffic in and out of the Minneapolis area is of such density that Center can't accept your request:

Ctr: *Cherokee One Four Six One Tango, unable at this time. Suggest you monitor this frequency.*

You: Roger, Center. Understand. Cherokee Six One Tango.

With only 80 miles or so to go, this isn't much of a problem. However, it does mean that the need for constant sky-scanning is more important than ever. If Center is too busy to give you advisories, you can be reasonably certain that there's a fair amount of activity in the surrounding area. Maximum alertness is in order.

Approaching Mason City, note that the airport has a Control zone but no tower. The airport used to have an FSS on the field, but when FSSs were consolidated, Mason City was left with only an independent weather observer (for commuter airline operations), an RCO tied into the Fort Dodge AFSS on 122.6, and UNICOM on 123.0.

So, after verifying that the CTAF is 123.0, you dial that in:

You: Mason City UNICOM, Cherokee One Four Six One Tango is over Hanontown at seven thousand five hundred for landing. Request airport advisory.

UNI: *Cherokee One Four Six One Tango, Mason City UNICOM. Wind two zero zero at one zero. Favored runway is One Seven. No reported traffic.*

You: Roger, Cherokee Six One Tango.

From this point on, stay on 123.0, but address your calls to "Mason City Traffic," not "UNICOM." In a few minutes you make your first traffic call:

You: Mason City Traffic, Cherokee One Four Six One Tango, five miles north at three thousand, descending for straight-in approach, Runway One Seven, full stop, Mason City.

You: Mason City Traffic, Cherokee Six One Tango on two-mile final for Runway One Seven, full stop, Mason City.

When clear of the runway.

You: Mason City traffic, Cherokee Six One Tango, clear of Runway One Seven taxiing to the terminal, Mason City.

At the ramp you decide to close the flight plan by radio rather than by telephone. Using the RCO you call the AFSS:

You: Fort Dodge Radio, Cherokee One Four Six One Tango on one two two pint six, Mason City.

FSS: *Cherokee One Four Six One Tango, Fort Dodge Radio, go ahead.*

You: Cherokee Six One Tango is on the ramp at Mason City. Would you close out my VFR flight plan from Minneapolis at this time?

FSS: *Cherokee Six One Tango, roger. Will close out your flight plan at five five.*

You: Thank you. Cherokee Six One Tango.

Mason City to Newton

The flight plan to Newton, 90 miles away, has been filed, you know the weather, and it's departure time again. Tune to 123.0 again and announce your initial intentions:

You: Mason City Traffic, Cherokee One Four Six One Tango at the terminal, taxiing to Runway One Seven, Mason City.

After the pre-takeoff check, with the second radio already turned to 122.6, open the flight plan:

You: Fort Dodge Radio, Cherokee One Four Six One Tango on one two two point six, Mason City.

FSS: *Cherokee One Four Six One Tango, Fort Dodge Radio, go ahead.*

You: Would you please open my flight plan to Newton at this time?

FSS: *Cherokee Six One Tango, roger. Will open your flight plan to Newton at two zero.*

You: Roger. Thank you. Cherokee Six One Tango.

Now go back to 123.0:

You: Mason City Traffic, Cherokee Six One Tango departing Runway One Seven, straight-out departure, Mason City.

When clear of the Control Zone:

> **You:** Mason City Traffic, Cherokee Six One Tango is clear of the area to the south, Mason City.

Following this call, you request Minneapolis Center to give you VFR advisories. Center in this location is remoted to Mason City on 127.3:

> **You:** Minneapolis Center, Cherokee One Four Six One Tango.
>
> *Ctr:* *Cherokee One Four Six One Tango, Minneapolis Center, go ahead.*
>
> **You:** Cherokee One Four Six One Tango is off Mason City at four thousand five hundred, climbing to seven thousand five hundred, en route Newton, and squawking one two zero zero. Request VFR advisories.
>
> *Ctr:* *Cherokee Six One Tango, squawk two five two five and ident.*
>
> **You:** Cherokee Six One Tango squawking two five two five.
>
> *Ctr:* *Cherokee Six One Tango, radar contact. Traffic at ten o'clock, four miles, southbound. Altitude unknown.*
>
> **You:** Cherokee Six One Tango is looking.

A minute or so later, you spot the target a little above you at the eleven o'clock position:

> **You:** Cherokee Six One Tango has the traffic.
>
> *Ctr:* *Cherokee Six One Tango, roger.*

When at your altitude:

> **You:** Cherokee Six One Tango level at seven thousand five hundred.
>
> *Ctr:* *Cherokee Six One Tango, roger.*

Very shortly, according to the Enroute Low Altitude Chart, you'll be leaving the airspace of Minneapolis Center and entering that controlled by Chicago. You can't be sure, but you'll probably be asked to change to the Des Moines remote outlet on 127.05. Assuming that that will be the frequency, dial it in so that you'll be prepared. After a few more minutes, Center comes on:

> *Ctr:* *Cherokee Six One Tango, contact Chicago Center now on one two seven point zero five. Good day.*

You:	Roger, one two seven point zero five. Thank you for your help. Cherokee Six One Tango.
You:	Chicago Center, Cherokee One Four Six One Tango is with you, level at seven thousand five hundred, en route Newton.
Ctr:	*Cherokee Six One Tango, roger. Des Moines altimeter two niner niner eight.*

Nearing Newton, you'll hear something like this:

Ctr:	*Cherokee Six One Tango, position one five miles north of the Newton VOR. Radar service terminated. Squawk one two zero zero. Frequency change approved. Good day.*
You:	Roger. One two zero zero. Thank you for your help. Cherokee Six One Tango.

Newton is uncontrolled, with only UNICOM on 122.8. About ten miles out, you announce your presence:

You:	Newton UNICOM, Cherokee One Four Six One Tango is five north of Newton VOR. Request airport advisory.
Uni:	*Cherokee One Four Six One Tango, Newton UNICOM. Wind is two one zero at one five. Altimeter three zero one five. Runway One Three in use. Three Cessnas reported in the pattern.*
You:	Roger. Cherokee Six One Tango.
You:	Newton Traffic, Cherokee One Four Six One Tango, eight miles northwest at four thousand. Descending for straight-in approach, Runway One Three, full stop, Newton.

You continue to descend and get lined up with the runway. Just as you're about to announce your position on the three-mile final, one of the Cessnas comes on the air: "Cessna Six Four Foxtrot turning base, touch-and-go, One Three, Newton."

You spot the Cessna in his turn to base and realize that the two of you are going to have a fairly close final together if things keep on as they are. You have a choice: "fly" the final with a series of S-turns, or do a 360. Of the two, the latter seems the wiser move:

You:	Newton Traffic, Cherokee Six One Tango is on a three-mile final but will do a three sixty to give way to the Cessna on base, Newton.

The Cessna may or may not thank you for your courtesy. Regardless, you complete the maneuver and line up for One Three again:

You: Newton Traffic, Cherokee Six One Tango on three-mile final for Runway One Three, full stop, Newton.

When down and clear of the runway:

You: Newton Traffic, Cherokee Six One Tango clear of Runway One Three, taxiing to the terminal, Newton.

When in the terminal, don't forget to cancel the flight plan. At Newton, this has to be done by phone to the FSS in Fort Dodge on 1-800-WX-BRIEF.

Newton to Kansas City

With the remaining passenger dropped off and full fuel tanks, you're ready for the last leg back to Kansas City. But first comes another call to Flight Service for a weather check and flight plan filing. Local conditions are determined from the UNICOM operator (or review the winds, etc., yourself if he's out gassing an airplane).

The radio contacts, in sequence, will be to local traffic, then to Flight Service, which you won't be able to reach until you have some altitude, and finally to Des Moines Approach for VFR advisories south on Victor 13. By the time you reach the Des Moines ARSA, you'll be well above its 5000 foot ceiling, so you don't have to call Approach Control. You will be going through the outer area, however, which would reach an approximate altitude of 12,000 feet MSL. Although you've decided to monitor Approach just to get an idea of potential traffic, you don't feel, considering the weather conditions, that it is necessary to request advisories for the very few minutes you'll be in the outer area. You won't hesitate to change your mind, though, if the monitoring indicates the need for those advisories.

While you're still on the ramp at Newton:

You: Newton Traffic, Cherokee One Four Six One Tango at the terminal, taxiing to Runway One Three, Newton.

After engine runup:

You: Newton traffic, Cherokee Six One Tango departing on Runway One Three, southwest departure, Newton.

You're off the ground, and at 300 feet or so, you contact Fort Dodge Flight Service over the Newton VOR. You transmit to the FSS on 122.1 and receive from the FSS on the VOR frequency of 112.5.

FSS: *Cherokee One Four Six One Tango, Fort Dodge Radio, to ahead.*

You: Fort Dodge, Cherokee Six One Tango was off Newton at three five past the hour. Would you please open my flight plan to Kansas City Downtown?

FSS: *Cherokee Six One Tango, roger. We show you off Newton at three five and will activate your flight plan to Kansas City. Des Moines altimeter three zero zero one.*

You: Roger. Thank You. Cherokee Six One Tango.

Although there is no ATA at Newton, one more local call is in order:

You: Newton Traffic, Cherokee Six One Tango now clear of the area to the southwest, Newton.

Monitoring Des Moines Approach as you pass through the ARSA's outer area, you call Minneapolis Center when clear of the area.

You: Minneapolis Center, Cherokee One Four Six One Tango.

Ctr: *Cherokee One Four Six One Tango, Minneapolis Center, go ahead.*

You: Cherokee One Four Six One Tango is fifteen southwest of the Des Moines VOR at six thousand five hundred VFR to Kansas City Downtown via Victor One Three. Squawking one two zero zero. Request VFR advisories, if possible.

Ctr: *Cherokee Six One Tango, squawk zero five two three and ident.*

You: Cherokee Six One Tango squawking zero five two three.

Ctr: *Cherokee Six One Tango, radar contact. Report level at six thousand five hundred.*

Ctr: *Cherokee Six One Tango, roger.*

There may or may not be further advisories from Center, depending on traffic. Whichever the case, you're soon over the Des Moines VOR and heading outbound on Victor 13.

A Cross-Country: To Put It All Together

After passing the Lamoni VOR, 53 miles out of Des Moines, you leave the Minneapolis Center area and enter the Kansas City Center area. As you cross the line:

Ctr:	*Cherokee Six One Tango. Contact Kansas City Center now on one two seven point niner. Good day.*
You:	Roger. One two seven point niner. Cherokee Six One Tango. Good day.
You:	Kansas City Center, Cherokee One Four Six One Tango is with you at six thousand five hundred.
Ctr:	*Cherokee Six One Tango, roger. St. Joe altimeter two niner niner six.*
You:	Roger, Cherokee Six One Tango.

About now is the time to tune the second VOR to Kansas City on 112.6. With one VOR head tracking you outbound from Lamoni and the other inbound to Kansas City, you should be smack in the middle of V13.

When you're approximately due east of St. Joseph, Missouri, you spot some lightning not too far distant and just to the right of your course. This can be an omen of bad stuff, so you decide to check with Center. But wouldn't Center take it on itself to advise you of potential en route weather problems? Not likely. In fact, it's *un*likely. It's thus up to you to initiate the request for information:

You:	Center, Cherokee Six One Tango. Request.
Ctr:	*Cherokee Six One Tango, go ahead.*
You:	Cherokee Six One Tango has lightning at one o'clock. Will my present course keep me clear of the storms?
Ctr:	*Cherokee Six One Tango, affirmative. Scattered thunderstorms are moving northeast, but you should be past the area at your present ground speed.*
You:	Roger. Thank you. Cherokee Six One Tango.

On the other hand, if a storm/Cherokee encounter seems likely, Center might offer this advice: "Cherokee Six One Tango, the storm activity is moving due east. Suggest right heading of two seven zero past St. Joe to Topeka and come in behind the weather." Keep in mind that such a suggestion is not a command. You're VFR, and have freedom as well as options.

Assuming the weather is not going to be a factor, Center will call you as you near the Kansas City TCA:

Ctr:	*Cherokee Six One Tango, position five miles north of the TCA. Contact Kansas City Approach on one one niner point zero.*

You:	Roger. One one niner point zero. Thank you for your help. Cherokee Six One Tango.

Before contacting Approach, tune to 124.6 for the Kansas City Downtown ATIS. With the current information copied on your cross-country notes (or firmly in mind), switch to Approach:

You:	Kansas City Approach, Cherokee One Four Six One Tango is with you, level at six thousand five hundred with Information November.
App:	*Cherokee Six One Tango, cleared into the TCA, direct Kansas City VOR. Maintain six thousand five hundred.*
You:	Roger, cleared into the TCA and direct the VOR, maintain six thousand five hundred. Cherokee Six One Tango.

Approach sees you cross the VOR:

App:	*Cherokee Six One Tango, turn left heading one eight zero, descend and maintain four thousand.*
You:	Left to one eight zero, out of six thousand five hundred for four thousand. Cherokee Six One Tango.
You:	Cherokee Six One Tango level at four thousand.
App:	*Cherokee Six One Tango, roger.*
App:	*Cherokee Six One Tango, turn right heading two one zero. Descend and maintain three thousand. Report the airport in sight.*
You:	Right to two one zero. Out of four for three, and we'll report the airport. Cherokee Six One Tango.
You:	Cherokee Six One Tango level at three thousand, and we have the airport in sight.

At this point, Approach will turn you over to the tower:

App:	*Cherokee Six One Tango, roger. Radar service terminated. Contact Downtown Tower on one three three point three.*
You:	Roger—and thank you for your help. Cherokee Six One Tango.
You:	Downtown Tower, Cherokee One Four Six One Tango is with you, level at three thousand.
Twr:	*Cherokee Six One Tango, continue straight in for Runway One Niner. Sequence later.*
You:	Roger, straight in for One Niner, Cherokee Six One Tango.

Twr:	*Cherokee Six One Tango, you'll be number two to land following a Citation on base.*
You:	Roger. Cherokee Six One Tango has the Citation.
Twr:	*Cherokee Six One Tango, roger. Caution wake turbulence landing Citation. Cleared to land Runway One Niner.*
You:	Cleared to land, Cherokee Six One Tango.

You watch the Citation touch down. To plan your landing because of the jet wake, you'd like a current wind reading.

You:	Tower, Cherokee Six One Tango. Wind check.
Twr:	*Cherokee Six One Tango, wind two one zero at one five.*
You:	Cherokee Six One Tango.

Because you're coming in on 19, these winds should blow the wake to the left of the runway, so you decide to land on the right, or upwind, side. You do so without difficulty and complete the roll-out.

Twr:	*Cherokee Six One Tango, contact Ground point niner when clear.*
You:	Cherokee Six One Tango.
You:	Downtown Ground, Cherokee One Four Six One Tango clear of One Niner. Taxi to the Flying Service.
GC:	*Cherokee Six One Tango, taxi to the Flying Service.*

By radio when parked, or by phone, you close out the flight plan—and the trip concludes without incident.

CONCLUSION

The point of this cross-country was to illustrate the typical radio procedures when using Ground Control, Tower, Approach/Departure, Center, Flight Service, and when operating within TCAs and ARSAs, and at UNICOM airports. The trip thus tried to encapsulate the more common phrases and phraseologies we discussed in the various previous chapters.

Yes, there were several instances of what you might have considered needless repetition, but a cross-country involves repetition. Many of the same things are said to different agencies. Besides, repetition has a way of cementing habits in one's mind and preventing errors and misunderstandings. Not every possible contact was included (as, for example, no call to Flight Watch), but many of the most common dialogues were recreated. And we hope that the use of the radio and the services available to every VFR pilot are just a bit clearer, and perhaps some of the radio technique concerns felt by so many VFR pilots have been allayed.

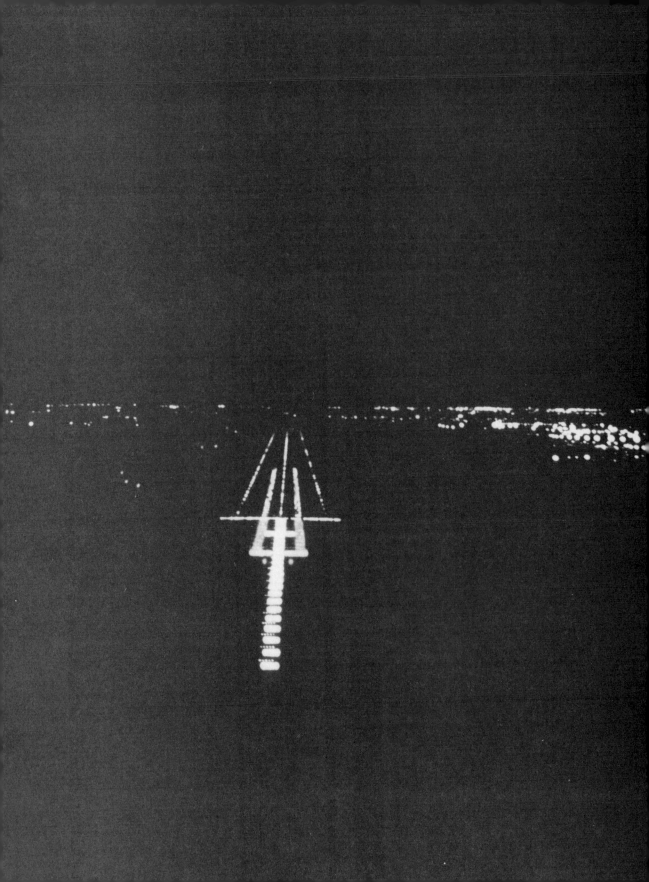

14

A Final Word

Proper radio procedures are perhaps the most overlooked or under-emphasized aspect in pilot training programs. Many budding pilots are never taught how to record and organize nav and com frequencies for easy reference and how to set up the frequencies in advance (assuming two navcoms are on board) to avoid excessive dialing at changeover points. They receive only the very basics of air-to-ground communications: what to say to whom and what to expect to hear in response. However, just as important as knowing *what* to say is *how* to say it.

Of course, there are exceptions to these criticisms. Some instructors like to teach communications and produce excellent pilots in the process. Their students sally forth into controlled areas with skill and confidence.

For those not privy to such training, it's a different story. Even among many "experienced" pilots, there is a fear that an instruction from a controller won't be understood and a fear of sounding stupid over the air. The result is that those pilots avoid controlled airports and other FAA services available to them—services already paid for by their tax dollars. The fears are both natural and understandable for pilots who haven't been trained in radio techniques.

Every one of us had qualms when we made those first tentative calls. Every one of us has since screwed up a transmission at one time or another. So what? Mistakes in flying an airplane can be very fatal; mistakes over the air are usually no more than embarrassing—if that. If you learn what to say, how to say it, and aren't afraid to ask a controller to "say again" or "say more slowly" when you haven't understood, you'll find that your flights, local or otherwise, will be safer and more secure, and you'll have the confidence to venture into that tower-controlled airport that you have perhaps been avoiding.

If this book has clarified just a few of the areas in the radio communications process, it has met its objective. We undertook the project because we felt the need existed for a work of this nature. By so doing, we hope that any concerns have been put to rest and that your flying will be more enjoyable.

There are others in the air, however, who have no concerns and no doubts, but are languishing in unrecognized ignorance. They bust into TCAs and ARSAs unannounced; they monopolize the air with trivia; they hem, haw, mumble, and meander; their messages are unplanned, their transmissions disorganized. They take four minutes to communicate what the pro does in four seconds. Are they aware of their deficiencies? You know better. The air is for them, and let the rest take the hindmost. As is so often the case, those who need help the most are the last to recognize that need and ask for the help.

The sky is for all of us. For those who like to venture forth, let's do it with confidence and professionalism. Radio skills don't make us better pilots at the controls, but they certainly add to our competence and, quite logically, our professionalism. Therein lies much of the joy of flying.

Abbreviations

The following is a glossary of various terms cited in this book. For a more complete glossary of pilot/controller radio communications terms, refer to the *Airman's Information Manual*, available by subscription from the U.S. Government Printing Office. Annual reprints are available from TAB BOOKS Inc., Blue Ridge Summit, PA 17294.

AAS: Airport Advisory Service
ADF: Automatic Direction Finder
A/FD: Airport/Facility Directory
AFSS: Automated Flight Service Station
AGL: Above Ground Level
AIM: Airman's Information Manual
AIRMET: Airman's Meteorological Information
App: Approach Control
ARSA: Airport Radar Service Area
ARTCC: Air Route Traffic Control Center
ATA: Airport Traffic Area
ATC: Air Traffic Control
ATCRBS: Air Traffic Control Radar Beacon System
ATIS: Automatic Terminal Information Service

CAVU: Ceiling And Visibility Unlimited
CD: Clearance Delivery
Center: Air Route Traffic Control Center (See ARTCC)
Com: Communications (Also, send-and-receive side of the radio [navcom])
CT: Control Tower

CTAF: Common Traffic Advisory Frequency
CZ: Control Zone

Dep: Departure Control
DF: Direction Finding
DF Fix: Direction Finding Fix
DF Steer: Direction Finding Steer
DME: Distance Measuring Equipment

EFAS: EnRroute Flight Advisory Service
ETA: Estimated Time of Arrival
ETD: Estimated Time of Departure
ETE: Estimated Time EnRoute

FARs: Federal Aviation Regulations
FSS: Flight Service Station

GC: Ground Control

HIWAS: Hazardous Inflight Weather Advisory Service

IFR: Instrument Flight Rules
IMC: Instrument Meteorological Conditions
IPAIDS: Identification-Position-Altitude-Intentions or Destination-Squawk

MEA: Minimum EnRoute Altitude
MOA: Military Operations Area
MOCA: Minimum Obstruction Clearance Altitude
MRA: Minimum Reception Area
MSL: Mean Sea Level
MULTICOM: Non-government air/air radio communication facility

Nav: Navigation (Also the navigation side of the radio [navcom])
Navcom: Radio with navigation and communication capabilities
NDB: Non-Directional Beacon
NM: Nautical miles
NOTAM: Notice to Airmen

PIREP: Pilot Weather Report
PTC: Pre-Taxi Clearance Delivery

RCO: Remote Communications Outlet

SIGMET: Significant Meteorological Information

SM: Statute miles

Squawk: Activate specific number code in the transponder

Stage III: Radar service that provides vectoring and sequencing for arriving VFR aircraft.

SVFR: Special Visual Flight Rules or Operations

TAC: Terminal Area Chart

TCA: Terminal Control Area

TRSA: Terminal Radar Service Area

TWEB: Transcribed Weather Broadcast

UHF: Ultra High Frequency

UNICOM: Non-government air/ground radio communication facility

VFR: Visual Flight Rules

VHF: Very High Frequency

VOR: Very High Frequency Omnidirectional Range Station (which provides course guidance information)

VORTAC: A VOR station combined with the military TACAN distance information (which provides course guidance plus nautical mile distance to the VORTAC station)

Wilco: Will comply